To The Aged P's
Happy 40th

KV-579-241

✶ INSIGHT POCKET GUIDE

ITALIAN Lakes

Discovery CHANNEL

APA PUBLICATIONS
Part of the Langenscheidt Publishing Group

L

Genève
Mt. Blanc
4808
SWITZERLAND
Chur
Innsbruck
AUSTRIA
Graz
Brig
A L P S
Bolzano
Klagenfurt
HUNGARY
Aosta
M. Rosa
4634
Varese
Trento
Pordenone
Udine
SLOVENIA
Maribor
FRANCE
Novara
Milano
Verona
Ljubljana
Zagreb
Torino
Trieste
Cuneo
Parma
Venezia
Rijeka
CROATIA
Genova
Ferrara
Pula
Nice
Modena
Bologna
Ravenna
Bihac
Banja
Luka
MONACO
La Spezia
Pisa
Firenze
Rimini
SAN MARINO
Zadar
BOSNIA
HERZEGOVINA
LIGURIAN
SEA
Livorno
Siena
Arezzo
Ancona
Split
Mostar
Bastia
I. d' Elba
Perugia
Corse
(Corsica)
Grosseto
Terni
Teramo
ADRIATIC
SEA
Dubrovnik
Ajaccio
Pescara
Porto
Torres
Bonifacio
Civitavecchia
Vasto
Olbia
ROMA
Campobasso
Foggia
Sassari
Latina
Oristano
Sardegna
(Sardinia)
Benevento
Bari
NAPOLI
Salerno
Brindisi
Sant'
Antioco
Cagliari
Vesuvius
1277
Matera
Taranto
Lecce
TYRRHENIAN
SEA
Agropoli
Castrovillari
Cosenza
Crotone
Isole Lipari
Vibo
Valentia
Catanzaro
Bizerte
Trapani
Palermo
Messina
Reggio
di Calabria
L'Ariana
Marsala
Sicilia
(Sicily)
M. Etna
3340
Tunis
Agrigento
Catania
IONIAN
SEA
TUNISIA
Nabeul
Gela
Siracusa
Kairouan
Sousse
MALTA
Valletta
Sfax
I. di Lampedusa
MEDITERRANEAN SEA
Italy
I. Kerkenah
60 km / 100 miles

Welcome

T his guidebook combines the interests and enthusiasms of two of the world's best-known information providers: Insight Guides, who have set the standard for visual travel guides since 1970, and Discovery Channel, the world's premier source of non-fiction television programming. Its aim is to bring you the best of the Italian Lakes in a series of tailor-made itineraries put together by Insight Guides' expert on Northern Italy, Lisa Gerard-Sharp.

There are 16 itineraries altogether, based upon Lake Maggiore (itineraries 1–5, including taking in the smaller lakes of Orta, Varese and Lugano), Lake Como (6–9), Lake Iseo (10–12) and Lake Garda (13–15). The final itinerary is an excursion to the city of Milan (which can be easily reached by train from any of the main lake towns) to see its soaring cathedral and shop in its exclusive shops. Each itinerary includes detailed instructions on transport, and ideas on where to lunch or dine on the way. Supporting the itineraries are sections on history and culture, shopping, eating out, sport, children's activities, a calendar of special events, and a fact-packed practical information section that includes a list of hand-picked hotels at all price levels

Lisa Gerard-sharp, the author of this guide, is a writer and broadcaster with a special interest in Italy. She has contributed to most of Insight's Italian titles, including *Insight Guide: Northern Italy*. Journalistic assignments for RAI TV in Rome provided her with a suitable pretext for forays to Lake Como, where soporific steamer crossings to sumptuous villas and gardens, faded hotels and fish dinners on the waterfront offered an escape from the Roman heat and the illusion of a slower pace of life. Since then, commissions to write features on fashion, food and wine for the British press have provided an excuse for full immersion in the lesser-known lakes, such as Orta, Iseo and Varese, as well as alpine scenery embracing hilltop villages, medieval castles and prehistoric rock carvings.

6   **contents**

contents

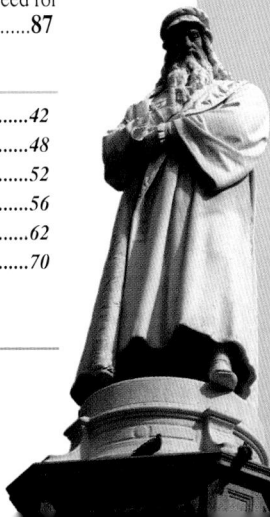

Pages 2/3: view of Como from the road to Bellagio
Pages 8/9: ferry coming into dock, Bellagio

History & *Culture*

H enry James's heart soared as he left Switzerland for the Italian lakes: 'On, on into Italy we went – a rapturous progress through a wild luxuriance of corn and olives and figs and mulberries and chestnuts and frescoed villages and clamorous beggars and all the good old Italianisms of tradition.' The sight of sluggish steamers and snow-clad peaks still stirs visitors. But, as James said of Lake Como, 'It's the place to enjoy *à deux* – it's a shame to be here in gross melancholy solitude.'

Lake Como is the lake most closely associated with the *fin-de-siècle* balls attended by grand dukes and dowager empresses, before democracy swept away the aristocracy. Not that Victorian-style promenading has ever gone out of fashion in the lakes. Parasols no longer twirl in the breeze but the romance remains as tangible as the scent of lemons or the taste of pistachio ice cream.

The lakes region's complex culture can be divided into discrete mini-cultures. The area covers Lombardy, Piedmont, the Veneto and Trentino, all of which have clearly separate identities. But it is the bond with, and loyalty to, their home town that many Italians feel most deeply. For instance *Comaschi* (Como natives) live in close proximity to Milan, one of Europe's great cities and yet, far from embracing the urban, cosmopolitan life, they identify only with their own town. In Como it is not only the glorious lakeside views and general tranquillity that enhance the wonderful quality of life: there are any number of designer shops and good restaurants.

Members of Como's old commercial families have known each other since infancy and do business together in a relaxed way. Their commitment to local culture embraces everything from the local silk industry to an appreciation of the medieval craftsmen who built Como's Romanesque churches.

The Lombard plain was hardly ideal terrain on which to settle: the marshes needed to be drained, the water channelled into canals. Yet by medieval times, it was domesticated and dotted with castles, churches, abbeys and palaces. This transformation, from bog to economic powerhouse, was due largely to the industriousness of the lakeside towns' natives and the advent of a prosperous local mercantile class.

Lombard Bankers

Lombard bankers lent money to the rest of Europe and invested in the lucrative arms and textiles industries. Italy's arms industry has been based in Brescia since the Middle Ages, with the chunky *Beretta* pistols used by today's police force merely a continuation of a process that began with swords, suits of armour and cannon balls. The region is also heir to an ancient crafts and textiles tradition and continues to churn out textile tycoons. The *Missoni* knitwear clan hails from

Left: medieval master builders. **Right:** Brescia's arms industry began with the manufacture of armour

Varese; the *Trussardi* fashion dynasty was born and bred in Bergamo. Lombardy's textile industry, with silk based in Como, flourished in the 16th century and it remains part of the fabric of local society.

Oetzi the Iceman

The remains of the earliest known northern Italian human being were unearthed in 1991 north of Trentino, where the discovery of a prehistoric cadaver continues to intrigue scientists. Christened Oetzi the Iceman, this 5,300-year-old corpse is one of the oldest and best-preserved Neolithic finds in Europe. The Stone Age body was recovered almost intact on an alpine glacier, complete with a bearskin hat, clothes made from goat and deerskin, and an armoury of arrows, longbow and copper axe. Much to the scientists' surprise, original tests indicated that Oetzi had been a strict vegetarian and died of natural causes, an unusual feat for a Stone Age hunter, leader and copper smelter. Now however, the consensus is that Oetzi not only dined on ibex (mountain goat) but that he died a violent death. Neolithic villages have been discovered close to Lake Garda and Lake Maggiore while in Valle Camonica, north of Lake Iseo, inscriptions carved into the rocks testify to the existence of a sophisticated Bronze Age civilisation that survived into Roman times.

The lakes, set amid the southern foothills of the Alps, have long cast a spell over visitors. The Romans were enamoured of the lakeside spas, which were praised by Virgil and Pliny; Catullus is reputed to have chosen Sirmione, on Lake Garda, as the site for his favourite villa. This delightful romantic spot, which was known as 'the pearl of peninsulas', still features the ruins of the Roman villa where the Veronese poet is said to have died. Beyond Sirmione, which was essentially a spa resort, the Romans established colonies in such strategic spots as Milan, Como, Brescia and Verona. The latter's Roman amphitheatre, one of the finest in the country, is the setting for the renowned opera festival. Brescia's Roman legacy includes a Capitoline temple, a forum, and a city museum full of busts, statuary and sophisticated mosaics from ruined lakeside villas.

In 568–9, the Lombards, who were a barbarian people from northern Germany, marched south and conquered the Po Valley and the lakes, making Pavia and Brescia their chief bases. Although this warlike people intermarried with the natives and espoused Christianity, they retained their industrious, militaristic, somewhat dour temperament. Some argue that it was the Lombards' 'iron in the

Above: Oetzi the 5,300-year-old corpse
Right: Roman amphitheatre at Verona

soul' that paved the way for the region's phenomenal industrial success. This murky period in history, encompassing the Lombard and Carolingian conquests, is well-presented in Brescia's stunning Santa Giulia museum.

In the north, the growth of merchant capitalism and the rise of the nascent middle classes engendered a progressive approach to government and self-rule. In particular, the mercantile Lombards developed the idea of the *comune*, an independent city or city state, which represented an evolutionary leap from feudalism. The *comuni* set great store by citizenship and civic pride, a mind-set which in turn fuelled political and cultural competitiveness with other cities. For several centuries these city states pursued an independent commercial policy and wielded influence out of proportion to their size.

Civic rivalry was channelled into great monuments and works of art, including castles, churches and palaces. In the former Roman colonies of Milan, Como, Bergamo, Brescia, Verona, the imprint of Roman architecture was re-interpreted for a new age as Romanesque. Como began to produce a school of master-builders who were the equal of any in medieval Europe. These stonecutters, masons and master-craftsmen, known as *Maestri Comacini,* were responsible for building or embellishing the Romanesque masterpieces around the lakes. The master-builders have left their mark in Como's churches and cathedral, and in the lakeshore's slender bell towers.

These shores, which constitute the border between the Mediterranean and northern Europe, are supreme castle country – most of the medieval lakeside ports have a defensive fortress. Verona and Brescia fought over Desenzano, which had the largest port on Lake Garda, until 1400, when the Republic of Venice threw its weight into the conquest of this lucrative trading post.

The Quest for Gracious Living

Two of the region's most impressive medieval castles – the moated Rocca Scaligera in Sirmione and the lofty Rocca Borromeo in Angera on Lake Maggiore – date from this period. In time, the need for civic defence became secondary to a trait still familiar in Italy: the quest for gracious living. Castello di Bornato, set among vineyards close to Lake Iseo, is a charming combination of castle and Renaissance villa. The area's soft, smoky, vine-clad landscapes found their way into the subtle backdrops of Leonardo da Vinci's paintings.

In the early 15th century, a period that has left Brescia with graceful Renaissance squares and a Venetian astronomical clock-tower, Venice ruled a number of Italian cities. The Venetians fortified their trading posts on Lake Garda, including the walled port of Lazise. Most seductive of all is the Venetian influence on Bergamo, from the Gothic windows to the heraldic lion, the symbol of the Venetian republic.

In the 16th century Spain triumphed over France in the battle for Italy.

Right: medieval castles were transformed into vine-clad Renaissance villas

In the 18th century Austria made significant gains in Italy and, for a brief but radical period, parts of the country were subjected to the rule of Napoleon.

Those living in the lakes region may have had little control over their lives, but there were, at least for the elite, compensations. Forced out of public life by a succession of foreign rulers, the locals took refuge in the melancholic loveliness of the lakes, around which they built sumptuous villas. Particularly in the Renaissance and baroque eras, patrician villa owners eagerly embellished nature. Employing sculptors, painters and landscape gardeners, they created airy loggias and Mannerist frescoes, parterres and pergolas. The Renaissance aesthetic of restraint and the baroque taste for theatricality furnished the villas with lakeside vistas framed by topiary and terraced gardens. Two of the finest villas are Villa Cicogna near Varese (a Lombard Renaissance villa set in Italianate gardens) and, on Lake Como, featuring the most romantic lakeside setting, Villa Balbaniello.

The baroque period (late 16th to early 18th century) was a golden era for construction: ostentatious villas and gardens embodied the aspirations of ambitious owners. With its landscaped gardens hugging the hill, Villa Carlotta on Lake Como exudes panache; Lake Maggiore's Isola Bella, its grounds full of statuary, fountains and grottoes, is a triumph of lofty terraces.

Habsburg Oppression

The transfer of power in Lombardy from Spain to Austria in 1714 revived Milan and fostered the spirit of the European Enlightenment. But before long

the locals began to resist Habsburg oppression. The *Risorgimento* was a 19th-century patriotic movement driven by Piedmont and its ruling House of Savoy that turned the lakes into a theatre of war. In 1858 the Piedmontese prime minister, Camillo di Cavour, persuaded the French to declare war on Austria. Subsequent French victories at Magenta

Above: a Venetian heraldic lion
Left: Villa Cicogna near Varese

and Solferino paved the way for the process of Italian unification. It was as a direct result of the particularly brutal Battle of Solferino that an appalled Jean Henri Dunant, a Swiss citizen, founded the Red Cross. Today Solferino has a commemorative museum, military fort and ossuary.

The Dolomites north of Lake Garda, controlled by the Austro-Hungarians from 1815 until 1918, was a front line in World War I. Italy joined the Allies in 1915, and soon made territorial gains, including Trentino, to the north. The Italian War Museum, set in a 15th-century Venetian-style castle in Rovereto, north of Lake Garda, charts the history of Italian warfare. In Gardone, on the lake, stands the bizarre home of the poet and patriot Gabriele d'Annunzio, who participated in raids on Bucharest and Vienna. He also captured a Dalmatian port. His villa, a testament to this heroic folly, includes the ship and planes used in his final fiasco, as well as indications of his fascist sympathies.

The disillusionment that followed the 'mutilated peace' of 1918 gave rise to Mussolini's Fascist party. Under *Il Duce*, Italy supported the Nazi conquest of Europe. As the Anglo-American forces made their way up through the peninsula in 1943, King Victor Emmanuel had Mussolini imprisoned. He was sprung from jail by the Nazis, who installed him in a puppet state known as the Republic of Salò. From 1943 to 1945, Villa Feltrinelli in Gargnano was the seat of the puppet government. Mussolini and his wife, Donna Rachele, were housed in a second, more secluded Villa Feltrinelli.

Meanwhile Gardone's Villa Fiordaliso was designed as a love nest for *Il Duce* and his devoted mistress, Claretta Petacci. Today, the official Villa Feltrinelli is associated with Milan University and Villa Fiordaliso is a prestigious *Art Nouveau* hotel still adorned by swastikas. As for Salò, its notorious decadence served only to swell the ranks of the Italian partisans. In April 1945, partisans caught Mussolini and Clara fleeing for their life at Dongo, on Lake Como. The couple were shot and later strung up by their heels with piano wire in Milan.

Tourism Takes Off

Blessed by a mild climate, romantic waterfront views and lush vegetation, the Italian lakes have long been a favoured haunt of the European elite. The German Romantic poets waxed lyrical about the lakes while composers such as Verdi, Rossini and Bellini sought inspiration on these shores.

In the 19th century, John Ball of the London Alpine Society climbed a mountain in the Dolomites and came down with the genesis of a tourist industry. British and Austrian mountaineers charted much of the Alps, and their journals became required reading for travellers to the lakes. In 1879 Queen Victoria visited Lake Maggiore, staying in Villa Clara at Baveno.

Above: the Battle of Solferino was brutal even by the standards of Italian wars
Next Page Top: Mussolini's heyday. **Bottom:** the resort of Limone sul Garda

In the 1870s, Lake Garda, and especially the resort of Arco, became a winter watering hole for Austrian grand-dukes, including Emperor Franz-Joseph and his cousin Albert. The resort was characterised by *belle époque* balls, health cures and carriage rides. Francesco II, former king of Naples and the Two Sicilies, died at Arco in 1894. The outbreak of World War I ended the lake's heyday.

While the aristocracy of *Mitteleuropa* flocked to the sanatoria around Lake Garda, the lake was also appreciated by writers and politicians: the Grand Hotel in Gardone Riviera was patronised by Vladimir Nabokov and Somerset Maugham, and it became Winston Churchill's base for painting holidays. D.H. Lawrence adored Limone sul Garda which, he wrote, overlooked 'a lake as beautiful as the beginning of creation'. Johann Wolfgang von Goethe and Henrik Ibsen concurred, though Goethe suffered an untoward experience at Malcesine, a delightful medieval town on Lake Garda. While he was sketching the castle Austrian police arrested him on suspicion of being a spy.

The Highest Life Expectancy in Europe

Italians are proud of spas such as Comano, Levico and Boario Terme, and of the lake's beneficial effects on 'the stressed and neurotic, the arthritic and the asthmatic, the elderly and young children'. Limone sul Garda has the highest life expectancy in Europe, and the greatest number of healthy citizens over 80. The absence of heart disease here has been studied by numerous scientists, who variously ascribe the region's health to the climate, the diet or genes.

The village was isolated until the 1930s, and only accessible by boat, so the secret may lie in the limited gene pool, which fostered the creation of antibodies and a rare blood group. A protein in the locals' blood, known as Apolipoprotein A-1, seemingly purges fat from the arteries. Beyond genes, a mild climate, a stress-free way of life and a cholesterol-free diet of lemons, lake fish and olives all contribute to the fine health of Lake Garda's residents.

HISTORICAL HIGHLIGHTS

10,000 BC Emergence of the lakes' Val Camonica civilisation (which survived into Roman times), with the first rocks carved by Camuni hunter gatherers.

AD 313 Emperor Constantine grants freedom of worship to Christians in Milan *(Mediolanum)*. Christianity is declared the official religion of the Roman empire.

4th century Milan becomes the *de facto* capital of the western Roman empire.

568–9 The Lombards, who occupy the Po Valley as far as the lakes, establish their capital at Pavia.

773–4 Charlemagne takes the Lombard crown.

800 Charlemagne, proclaimed emperor by the Pope, commands a union of western Europe.

1024 Milan's first popular assembly *(parlamento)* marks the emergence of the *comuni*, or independent city states.

1152 The German prince Frederick Barbarossa is named Holy Roman Emperor.

1347–8 The Black Death devastates the population of northern Italy.

1386 Gian Galeazzo Visconti, ruler of Milan, initiates the building of Milan Cathedral.

1405 Venetians conquer Verona, Padua and Bergamo.

1494 The French victory at the Battle of Fornovo marks the beginning of several centuries of foreign rule over northern Italy.

1530 Charles V is crowned Holy Roman Emperor and Lombardy comes under his rule.

1545 The first Council of Trent highlights the Counter Reformation's attempt to wrest the initiative back from Protestantism.

1714 Lombardy and much of the lakes pass from Spanish rule to that of the Austrian (Habsburg) empire.

1796 Napoleon invades northern Italy.

1815 Restoration of Habsburg rule in northern Italy.

1848 People of Milan rebel against the Habsburgs.

1859 France defeats Austria at the Battle of Solferino. Milan joins the Kingdom of Italy.

1861 Victor Emmanuel II (Vittorio Emanuele) is declared the first king of Italy.

1866 Giuseppe Garibaldi leads a volunteer army in a new war against Austria, after which Venice becomes part of Italy.

1871 Italian unification.

1915 Italy joins the World War I Allies.

1922 Fascists under Benito Mussolini seize power in Italy.

1940 Italy enters World War II as an ally of Nazi Germany.

1943–5 Italy surrenders to the Allies, Mussolini is installed in the Republic of Salò until his capture and execution near Como.

1951 Italy joins the forerunner of the European Union as a founder.

1987 Italy wins a place among the G7 top industrialised nations.

1992 *Tangentopoli* ('Bribesville') corruption scandals rock the north and lead to an overhaul of public life.

1994 Emergence of *Forza Italia*, a new right-wing political party led by media tycoon Silvio Berlusconi, who serves briefly as prime minister.

2000 The redesigned Malpensa airport becomes a gateway to the lakes, as Verona, Brescia and Bergamo become increasingly popular.

2001–present Silvio Berlusconi's second shot at the premiership proves controversial, with issues over his personal probity and stranglehold over the media.

2002 Introduction of the Euro currency, which takes the place of the lira.

2006 The Winter Olympics are scheduled to be held in Turin.

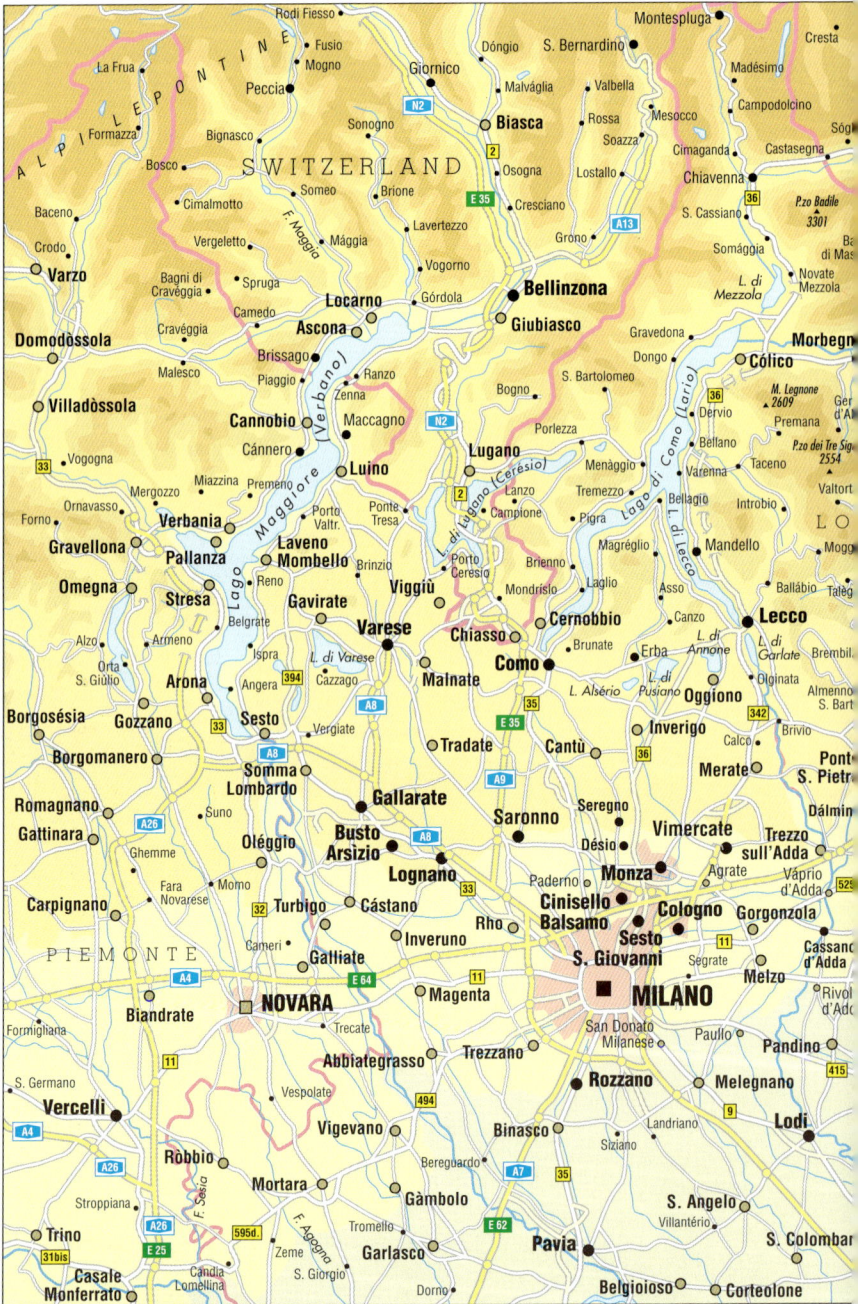

Rodi Fiesso
Fusio
Mogno
Giornico
Dóngio
S. Bernardino
Montespluga
Cresta
Madésimo
La Frua
Peccia
Malváglia
Valbella
Campodolcino
Sóg
ALPI LEPONTINE
Formazza
Bignasco
Sonogno
Biasca
Rossa
Soazza
Mesocco
Cimaganda
Castasegna
Bosco
SWITZERLAND
Someó
Brione
Osogna
Lostallo
Chiavenna
P.zo Badile
3301
Ba
di Mas
Baceno
Cimalmotto
F. Maggia
Lavertezzo
Cresciano
Grono
S. Cassiano
36
Crodo
Vergeletto
Mággia
Vogorno
Somággia
Novate
Mezzola
Varzo
Bagni di
Cravéggia
Spruga
Camedo
Górdola
Locarno
L. di
Mezzola
Domodòssola
Cravéggia
Ascona
Bellinzona
Gravedona
Morbegn
Malesco
Brissago
Piaggio
Ranzo
Giubiasco
Dongo
Cólico
Villadòssola
Cannobio
Zenna
S. Bartolomeo
M. Legnone
2609
Ger
d'A
36
Dervio
Premana
Vogogna
Cánnero
Maccagno
N2
Bogno
Porlezza
Lago di Como (Lario)
P.zo dei Tre Sig
2554
33
Miazzina
Premeno
Luino
Lugano (Ceresio)
Menàggio
Bellano
Varenna
Taceno
Valtort
Ornavasso
Mergozzo
Porto
Valtr.
Ponte
Tresa
Lanzo
Tremezzo
Bellagio
Introbio
Forno
Verbania
Laveno
Mombello
Brinzio
Porto
Ceresio
Campione
Pigra
Magréglio
L. di Lecco
Mandello
Mogg
Gravellona
Pallanza
Lago
Reno
Viggiù
Mondrislo
Brienno
Laglio
Asso
Ballábio
Taleg
Omegna
Stresa
Gavirate
Maggiore
Belgrate
Varese
Cernobbio
Canzo
Lecco
LO
Alzo
Armeno
Ispra
Chiasso
Brunate
Erba
L. di
Annone
L. di
Garlate
Brembil
Orta
S. Giúlio
Arona
Angera
Cazzago
L. di Varese
Como
L. Alsério
L. di
Pusiano
Oggiono
Olginata
Almenno
S. Bart.
Borgosésia
Gozzano
Sesto
394
Malnate
35
Inverigo
342
Calco
Brivio
Pont
S. Pietr
Borgomanero
A8
Vergiate
Tradate
Cantù
36
Merate
Romagnano
A26
Suno
A8
SOMMA
Lombardo
A9
Saronno
Seregno
Vimercate
Dálmin
Gattinara
Ghemme
Oléggio
Busto
Arsizio
A8
Désio
Trezzo
sull'Adda
Agrate
Váprio
d'Adda
52
Carpignano
Fara
Novarese
Momo
32
Turbigo
Cástano
Lognano
33
Rho
Paderno
Monza
Cinisello
Balsamo
Cologno
Gorgonzola
Cassano
d'Ad
PIEMONTE
Cameri
Galliate
Inveruno
Sesto
S. Giovanni
11
Segrate
Melzo
Rivol
d'Ad
A4
NOVARA
E 64
Magenta
11
MILANO
San Donato
Milanèse
Paullo
Pandino
415
Biandrate
Trecate
Abbiategrasso
Trezzano
Formigliana
11
Vespolate
Rozzano
Melegnano
S. Germano
494
Binasco
Landriano
9
Lodi
Vercellì
Vigevano
Siziano
A4
Bereguardo
A7
35
S. Angelo
S. Colombar
A26
Ròbbio
F. Sesia
Mortara
Gàmbolo
Villantério
Stroppiana
F. Agogna
Tromello
E 62
Trino
A26
595d.
Zeme
S. Giorgio
Garlasco
Pavia
Belgioioso
Corteolone
31bis
E 25
Candia
Lomellina
Dorno
Casale
Monferrato

Italian Lakes

20 km / 12 miles

SWITZERLAND

St. Moritz
Silvaplana
L. da Silvaplana
Sils
Lòbbia
Vicosoprano
Chiaréggio
M. Disgrázia 3678
Chiesa in Valmalenco
Cataéggio
Sondrio
Caiolo
Faedo
Ponte in Valtellina
Téglio

Bagni di Bormio
Bormio
Arnoga
la Rosa
Cima de Piazzi 3439
P.zo Bernina 4049
Bérnina
Poschiavo
Valbrutta
Caspóggio
Campocologno
Mad. di Tirano
Tirano

Ortles 3902
S. Catarina Valfurva
P.ta S. Matteo 3684
Pejo Terme
Temu
Ponte di Legno
Vezza d'Oglio
Edolo

Bagni di Rabbi
Malè
Mezzana
Vermiglio
Folgárida
Madonna di Campiglio
Carisolo
Cima Presanella 3556

Dímaro

TRENTINO ALTO ADIGE

Molveno
L. di Molveno

SÓNDALO
Grósio

Aprica
Malonno
Cédegolo
Saviore d. Adamello
Adamello 3554

Vigo Rendena
Tione di Trento
Breguzzo
Lardaro
Daone
Tiarno
Cimego
Condino
Storo
L. d'Idro (Eridio)
Anfo
Lavenone
Idro
L. di Valvestino
Tignale

Stènico
Ponte Arche
Comano Terme
Ballino
Dro
Tenno
Riva del Garda
Arco
Tórbole
Mori
L. di Ledro
Limone sul Garda
Magasa
Tremósine
S. Valentino
Ávio
Malcésine
Brenzone

OROBIE
P.zo di Coca 3050
Fóppolo
Mezzoldo
Carona
Piazzatorre
Valbondione
Lizzola
Schilpário
Dezzo
Borno
Bratto
Boàrio Terme
Bienno
Breno
Capo di Ponte
Bagolino
Pte. Cáffaro
Cóllio

ARDIA
S. Giovanni Bianco
Oltre il Colle
P.zo Arera 2512
Clusone
Rovetta
Serina
S. Pellegrino
Zogno
Gazzaniga
Albino
Villa d'Almè
Casazza
Alzano
Tavèrnola
Monte Isola
Predore
BERGAMO
Tréscore
Sárnico
L. d'Iseo (Sebino)
L. di Éndine
Darfo
Lòvere
Pisogne
Bóvegno
Gardone
Barghe
Lumezzane
Nave
Iseo
510
42
A4
Palazzolo sull'Oglio
E 64
Urgnano
Martinengo
573
Rovato
Ospitaletto
Romano
Chiari
Maclódio
Castenédolo
Caravàggio
Antegnate
Mozzánica
Orzinuovi
Pompiano
Bagnolo Mella
Ghedi
Soncino
Manérbio
Leno
Crema
Quinzano
Verolanuova
astelleone
Soresina
Pontevico
Casalbuttano
Robecco
Adda
Pizzighettone
Castiglione
415
Codogno
Cremona
Cicognolo
Pescarolo
Casalromano
Canneto

eviglio
Brozzo
Vestone
Gardone Riviera
Toscolano-Maderno
Gargnano
Pai
Salò
Gavardo
Manerba del Garda
Bedizzole
Rezzato
BRESCIA
45bis
Desenzano di Garda
Sirmione
Montichiari
Castiglione
Carpenédolo
Castel Goffredo
Isorella
Gámbara
Ásola
Casaloldo
Gazoldo
Acquanegra
45bis

Lago d'Idro (Eridio)
Lago di Garda (Benaco)
Ferrara di M. Baldo
F. Adige
Peri
S. Anna
Caprino Veronese
Torri d. Benaco
Garda
Bardolino
Cisano
Volgarne
Dolcè
A22
12
A4
Castelnuovo del Garda
11
Peschiera di Garda
S. Martino
E 70
450
Sommacampagna
62
Solferino
Volta
236
Valéggio
Villafranca di Verona
Gòito
Roverbella
Marmirolo
F. Mincio
A22
E 45
Mantova

VENETO

38
42
42
12
E 45
E 70
A21

F. Adda
Valtellina
ALPI
ALPI
RETICHE
F. Chiese
F. Oglio
F. Chiese

Orientation

Lake Maggiore, which borders Piedmont, Lombardy and the Swiss canton of Ticino, is Italy's second-largest lake. Stresa, the main resort on the more overtly spectacular Piedmont shore, is easily accessible from Milan by road or rail. Alternatively, the quietly appealing Lombard shore can be approached by steamer from Stresa, or by road via the provincial capital of Varese. For convenience, Stresa, followed by Baveno and Pallanza, are recommended; for serenity on the Lombard shore, choose Laveno or Angera.

Two lovely lakes lie west of Lake Maggiore: Lake Orta is an atmospheric spot noted for its spiritual air and excellent small hotels; Lake Mergozzo is an unspoilt place for walking and swimming. East of Lake Maggiore are the newly clean Lake Varese and Lake Lugano, which has a pretty stretch in Varese province. Both lakes are accessible from Varese, Como, Milan and Lugano, but they lack the charm of Maggiore and Como resorts.

Romantic Lake Como, north of Milan, is accessible from Bergamo, Monza, Milan and Malpensa airport. With its dramatic scenery and quaint ports, it has long appealed to British and American visitors, particularly independent travellers and couples. Bellagio is the most select resort, Como the most convenient, but Tremezzo, Menaggio and Cernobbio are also worth visiting.

Captivating Lake Iseo, which measures 25km (15 miles) by 5km (3 miles), encloses Europe's largest lake island, and features wild scenery on the western shore, as well as wine-growing hills to the south. The lake is accessible from Como, Bergamo and Brescia, with rail connections to Lake Garda.

Italy's Largest Lake

Garda, Italy's largest lake, is 48km (30 miles) long and up to 16km (10 miles) wide. The only major lake without a significant island, Garda offers fine beaches, gardens and ports, plus resorts, such as Bardolino, that are noted for their wine. The area also enjoys a wide range of climate conditions, from alpine glaciers just north of the lake to Mediterranean warmth – the lake acts as a kind of solar battery. Desenzano is the main rail link on the lake and there are also connections to Brescia, Verona and beyond. The most charming resorts are Riva and Limone in the north, Gardone and Salò in the west, Sirmione in the south, and Malcesine, Bardolino and Torri del Benaco in the east.

Given the conspicuous presence of the major tour operators on the lake, prices tend to be more competitive than on other lakes. Moreover, compared with the quiet refinement of Lake Como, for instance, Lake Garda features more in the way of family holidays and sports facilities.

Left: view of Rocca di Angera
Right: a ferry captain at the helm

Lake **Maggiore**

1. THE BORROMEAN ISLANDS *(see map below)*

An island-hopping day trip from the *belle époque* resort of Stresa to
Isola Bella, Isola Madre and Isola dei Pescatori. Explore these small,
exotic outcrops' villas and grounds, taking lunch in Isola Bella or Isola
Madre, and dinner in a rustic fish restaurant on Isola dei Pescatori.

*The Stresa ferry station on Piazza Marconi is next to the tourist office. Buy
a ticket that includes entrances to the island sights. A free private boat takes
diners to Stresa from the Verbano fish restaurant on Isola dei Pescatori.*

Located off the Stresa shore, the Borromean Islands bask in the warmth of
a Mediterranean micro-climate. Even though these lovely outcrops constitute
the most popular excursion in the lakes, they retain their languid charm.
Henry James described 'the delicious old Borromean Islands' as 'a quaint
mixture of tawdry flummery and genuine beauty, a sort of tropical half-

Lake Maggiore
5 km / 3 miles

Itinerary 1
Itinerary 2
Itinerary 3
Itinerary 4

splendid, half-slovenly Little Trianon and Hampton Court'. In those days tourists were rowed to the islands by hotel lackeys; today we depend on the bustling ferry crossing.

A Powerful Dynasty

Lake Maggiore was a fiefdom of the noble Borromei clan in the15th century, and the family has succeeded in holding onto some of the most beautiful parts, not least the Borromean Islands. This powerful Milanese dynasty has produced patrons of education, religious reformers, cardinals, popes and even a saint, San Carlo. The family scions are still fabulously wealthy owners of palaces in Milan and priceless art collections. These days they also receive the revenue from toll roads and fishing rights over the lake. The islands are strictly controlled fiefdoms; Isola Bella is one of the family's many private homes.

The quieter islands are easily visited in a day. Whereas Isola Bella was always intended to be a showy pleasure palace, Isola dei Pescatori was styled as a rural retreat, Isola Madre as an enchanted garden. In recent times **Isola Bella** has served as the Borromean princes' summer residence – the family stays in the island palace (open 9am–noon, 1.30–5.30pm) for several months of the year. To protect the family's privacy, and its art collection, two-thirds of the palace is closed to the public, but it is still worth seeing, not least for the beguiling baroque gardens and the palatial treasures on show.

When the current curator of the palace was appointed, he was taken aback by a clause in his contract requiring him to stay overnight on the island. While many visitors would jump at the opportunity, he was afraid of spending his first night alone on Isola Bella, facing ghosts or burglars drawn to the princely portraits. Now happily ensconced with his family, his fears have been allayed, even if he has to share the Tiepolo with hordes of tourists.

Although Isola Bella began as little more than a rock with a view, in the course of centuries it became a delightfully harmonious folly. In the 1620s, Count Carlo III Borromeo was inspired to create a full-blown monument to his wife, Isabella. To realise his vision, the rocks were transformed into an island with 10 terraces designed to resemble the prow of a ship in full sail. Boatloads of soil were transported to this barren island, as well as baroque statuary and the building materials needed for the creation of a palatial villa.

The works continued under his son, Vitiliano VI, and were virtually complete

Above: plying the calm waters
Right: refreshing greenery on a stone patio

lake maggiore

by the time of his death in 1670. Even so, inspired by the original plans, family descendants continued to embellish the island until the 1950s, when Vitiliano IX, the last member of the family to attempt major modifications, died before building his cherished harbour.

Somewhat ironically for a family whose motto celebrates humility, the turreted Borromean palace is a bombastic enterprise that reveals lamentable lapses in taste rather than the architect's lofty intentions. Beyond the cobbled courtyard, the palace's high-ceilinged rooms contain a fine collection of

16th- to 18th-century northern Italian art from Venice, Lombardy and Emilia. Unfortunately the art is encrusted in a baroque clutter of stucco-work and heraldic crests only partially redeemed by bold flourishes such as a cantilevered spiral staircase. Lavishly over-stuffed public rooms connect a gilded throne room, an empire-style ballroom and a Flemish long gallery featuring tapestries emblazoned with the unicorns that adorn the dynasty's distinctive crest.

Napoleon Slept Here

The palace has witnessed numerous momentous historical events and played host to emperors and statesmen. In 1797 Napoleon slept in one of the bedrooms – a ponderous neoclassical chamber decorated in what was the politically correct Directoire style. The ornate music room was the setting for the Stresa peace conference of 1935: it was here that Italy, Britain and France failed to agree on a strategic response to Hitler's programme of rapid re-armament, a missed opportunity that hastened the momentum towards war.

Designed as a cool summer retreat, the mysterious area beneath the palace conceals bizarre **artificial grottoes**, with tufa-stone walls studded with shells, pebbles and fossils. The maritime mood is sustained in the statue of a coolly reclining nude and caverns dotted with marine imagery. This grotesque creation reflects the contemporary taste for *wunderkammer* – chambers of marvels designed to enchant visitors with their eclectic displays. In the case of Isola Bella, however, the greatest marvel lies outside.

The **baroque gardens** envelop the palace in sweeping arcs, with dramatic architectural perspectives accentuated by grandiose urns, obelisks, fountains and statues. Although the terraced gardens abound in shady arbors, whimsical water features and mannered statuary, the sum is greater than its parts. Serried ranks of orange and lemon trees meet geometric flowerbeds before fading into a studied confusion of camellias and magnolias, laurels, cypresses, jasmine and pomegranate.

The ship-shaped terraces are crowned by

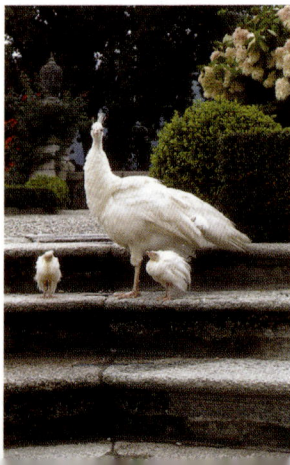

Above: Isola Bella's bizarre tufa-stoned grottoes
Right: white peacocks strut their stuff at Isola Bella

a four-tiered folly studded with shells and topped by cherubs and classical gods, as well as by the heraldic unicorn. The *pièce de résistance* is the shell-shaped amphitheatre that serves as a delightful setting for summer concerts. The rest of the small island is awash with trinket stands and poor quality tourist restaurants, so stay just long enough for lunch (possibly at Elvezia; tel: 0323-30043) before taking the first ferry to Isola Madre.

The Wildest Isle

Enjoy the cooling breezes on the ferry to **Isola Madre** before partaking of the quiet pleasures of these floating botanical gardens. Here on the largest, and what was the wildest of the islands, you will find gently landscaped gardens. The island is home to Europe's largest Kashmiri cypress and some of the first camellias planted in Italy. Best seen in spring, the camellias are part of a patchwork formed by shady paths, ancient cedars, mimosa, magnolias and giant rhododendrons. White peacocks, Chinese pheasants and parrots nest in the topiary or trail their tails over beds of budding camellia.

The 16th-century villa (9am–noon, 1.30–5.30pm) – an apparently austere Mannerist affair created as a home rather than a palace – is a bit stifling and gloomy. The Borromean family indulged their love of theatre here – check out the curious collection of puppet theatres, puppets and dolls.

Isola Pescatori, confusingly also known as Isola Superiore, is a pretty fishing village less dedicated to catching fish than to the preservation of its film-set picturesque appearance. It is a place for pottering down tiny alleys and peering at the lake, or even paddling off the pebble-beach overlooking Isola Bella. Adjoining the beach, Verbano (tel: 0323-30408) is the best restaurant on the Borromean Islands. You might want to return to this romantic place in the evening, courtesy of the (free) restaurant boat. Verbano's fish-based cuisine might be overrated but the lovely setting is the best in Stresa. A stroll through the maze of passageways conjures up the mood of a remote Greek island: a frescoed parish church fades into a view of fishing nets drying in the sun or courtyards of lithe cats basking on whitewashed walls. Henry James rightly praised this pocket of Italy for making one feel 'out of the rush and crush of the modern world'.

Above: welcome shade on Isola Pescatori

2. CRUISE FROM STRESA TO ANGERA CASTLE AND SANTA CATERINA *(see map, p22)*

From Stresa, take the cable car to the botanical gardens then catch a late-morning boat to Rocca d'Angera, the medieval castle that faces Stresa. After lunch in Angera, rejoin the cruise to Santa Caterina del Sasso, a 13th-century hermitage above the lake. Sail back to Stresa for a pre-prandial stroll around town. Dining options include a supper cruise.

Begin at Stresa tourist office on Piazza Marconi's jetty. Check sailing times with the tourist office (tel: 0323-30416) or the lake navigation service (free-phone: 800-551801) at least a day in advance. The classic Stresa-Angera-Santa Caterina cruise is usually a return trip that departs at around midday. Sometimes a later cruise on an old steamship visits Stresa, Arona and Angera. Santa Caterina is also served by hydrofoils from Arona. If taking the classic cruise, consider booking lunch in Angera (see Eating Out, page 76) and a dinner-dance or disco cruise from Stresa (phone numbers above).

Stresa, the main resort on the Piedmontese shore, was a *belle époque* wintering ground, the popularity of which was enhanced by the opening of the Simplon tunnel in 1906. A dowager resort distinctly past its prime, Stresa is considered to be the noble part of Lake Maggiore. It found favour with Queen Victoria and Winston Churchill, and remains popular with super-annuated politicians and former world leaders: Margaret Thatcher and Helmut Kohl are both known to have a soft spot for Stresa.

Heyday Hotels

Stresa's bustling waterfront at **Piazza Marconi** marks the start of the lakeside promenade to the cable-car station at the sandy lido, from where there are water taxis to the Borromean Islands. The 15-minute stroll towards Baveno passes Stresa's grandest hotels, the sole relic of its *fin-de-siècle* heyday. The 18th-century Villa Ducale, now an obscure museum, gives way to the Art Nouveau Regina Palace, and its renowned restaurant, followed by the Grand Hotel des Iles Borromees, which featured in Hemingway's *A Farewell to Arms*. Replete with beguiling vistas over the Borromean Islands, the lakeside public rose gardens have a sedate air that complements the resort's genteel reputation.

A **cable car** leads from the lido to **Monte Mottarone**, which commands views of the misty lakes, glacier-crested peaks and Monte Rosa ski slopes, as well as of the Domodossola valleys and the Lombard plain in the distance. (The 18-minute cable-car ride departs every 20 minutes from 9.20am, with a break for lunch. The last descent to Stresa is at 5pm). On a fine day there are bird's eye views of villas hidden in the pine groves. It is worth getting out at Alpino, the first stop, to see the exotic alpine rock gardens of **Giardino Alpino** (Tues–Sat 9.30am–6pm) before continuing to the top. Retrace your steps to Piazza Marconi to board the ferry to Angera.

Above: relaxing with a book on the promenade at Stresa. **Top Right:** a shady arbour in the Villa Pallavicino's gardens. **Right:** Rocca Borromeo

This **classic cruise** hugs the western shore, passing Villa Pallavicino, Belgirate and Meina before crossing to Angera, where the ferry docks, allowing time for lunch and a visit to the castle. This stretch of the western shore, all part of Piedmont, is home to some prestigious villas. Pallanza, north of Stresa, features the designer Alessi's villa, heralded by his trademark coffee pot, while Baveno's eclectic Villa Branca is known for its Fernet-Branca liqueur.

Supermodels and Footballers

On the outskirts of Stresa, the neoclassical Villa Pallavicino has landscaped grounds and an appealing zoo *(see Children, page 83)*. Further south, towards Arona, are clusters of celebrity villas, especially around **Meina**, where fashion designers such as Armani and Ferre have homes. You might catch tantalising glimpses of neoclassical mansions that are now home to British supermodels, and footballers such as Alessandro del Piero.

The ferry crosses to the Lombardy side of the lake and several of the finest sights in underrated Varese province. In **Angera**, the medieval hulk of **Rocca Borromeo** (9.30am–12.30pm, 2–6pm), the most imposing fortress on the lake, soon looms into view. The castle has belonged to the Borromeo dynasty since 1450. In the 14th century another Milanese *visconto* had covered the walls with frescoes celebrating the glory and longevity of his own dynasty. These fine frescoes are in the Gothic Sala di Giustizia, a vaulted chamber adorned with signs of the zodiac and scenes of military victories.

Elsewhere in the castle is a historical dolls' and children's costume museum. Displays include yellowing christening gowns, intricate court costumes and fat, grumpy-faced porcelain dolls that look more like miniature adults than childish toys. Rickety steps lead to the tower and views over vineyards and the lake.

For lunch, choose between the Rocca d'Angera in the castle courtyard or Hotel Lido *(see Eating Out, page 76)* on the lake. There should be views of **Arona**, facing Angera on the Piedmontese shore, which did have a castle,

lake maggiore

until its destruction by Napoleon. The Sancarlone – a monument to the locally born, beatified reformer San Carlo Borromeo (1538–84) – is the world's tallest statue after New York's Statue of Liberty.

From Angera, the ferry heads north to Santa Caterina, with a stop at the hermitage, passing an unspoilt stretch of coast in Varese province. The cruise passes **Ranco**, with its transport museum *(see Children's Activities, page 83)*, and **Ispra**, where there is an EU research centre. This institution was originally the *Euratom* centre for the study of atomic energy, but now focuses on raising European standards in industry and technology, with community projects such as the cleaning of Lake Maggiore. As evidence of the scientists' success, locals point to the presence of *lavarelli*, tiny, deep-water sprats that live off plankton, and thrive only in clean water.

Beyond Ispra lie the appealing port of **Monvalle** and the pipe-making centre of Brebbia. Geraniums, azaleas, begonias and petunias are cultivated in the

hinterland. On the lake, the silence is shattered only by the overhead whirring of *Agusta* helicopters' test flights. This side of Lake Maggiore is dismissed as the *sponda magra*, the 'poor shore', due to its lack of hotels.

Santa Caterina (9am–noon, 2–5pm; tel: 0332-647172), seemingly suspended over a rocky precipice and overhung by crags, is at its most romantic when viewed from the lake. Set on the only stretch of Lake Maggiore that has no banks, the hermitage guards the deepest parts of the lake. According to legend it was founded by Alberto Besozzi, a shipwrecked 12th-century money-lender who vowed to become a hermit if he survived. In 1195, his piety was said to be instrumental in averting a plague, and he was rewarded with the building of a hermitage to the 3rd-century St Catherine of Alexandria. A votive chapel was modelled on the monastery on Mount Sinai where the saint's body was supposedly borne by angels.

Prior to the dissolution of the monasteries in 1770, the sanctuary was variously Augustinian, Ambrosian and Carmelite. After centuries of decline, culminating in the collapse of the church roof in a landslide in 1910, it was saved by Varese province, which dedicated huge resources to its long-term restoration.

Gothic Frescoes

The monastic complex, strung out along the rocky ledge, is distinguished by a Gothic bell tower, Renaissance porch and an airy gallery overlooking the lake and the Borromean Islands. Gothic frescoes in the chapterhouse feature St Eligius healing a horse, and a later addition, the Carmelites' sacred emblem. Under the graceful Gothic loggia of the monastery is a *danse macabre*, complete with a very grim Grim Reaper. Other scenes depict the vanity of human aspirations: a merchant engrossed in his accounts and an amorous courtier face their own mortality.

Above: Santa Caterina

The church, essentially an amalgamation of all previous sanctuaries built on the site, is preceded by graceful frescoes depicting an intertwined trio of female saints, including St Catherine. Inside the frescoed sanctuary, but no longer visible and little more than a fissure in the rock, is Besozzi's cave.

Since 1975, the sanctuary has been entrusted to a small group of lay brothers led by a Benedictine monk. The cornerstone of community life is still work, meditation and prayer in a sanctuary suspended between lake and sky. Members of the community pursue their own interests – including bookbinding, antique-dealing, computing – interspersed with the restoration of church organs and the manufacture of the community's honey or its noxious Nocino liqueur. Religious retreats such as Santa Caterina do sometimes accept applications from like-minded lay people so, if you are interested, you could find yourself sipping Nocino for years to come.

The Best Ice Cream

Once the ferry returns to **Stresa**, which faces the sanctuary from the Piedmontese shore, you might want to regain your land legs before going to your hotel or embarking on a dinner cruise. The resort is short of specific sights but it does possess a certain faded charm, from a tiny harbour full of bobbing fishing boats, to the sunny **Piazza Cadorna**, the main inland square. Sit under the plane trees and sip a drink in Caffe Nazionale or get an ice cream at the neighbouring Angolo del Gelato, the town's best ice-cream parlour.

This is prime shopping time, so you might saunter down Via Bolognaro to see the old-fashioned shops or, more profitably, continue to Via Garibaldi, parallel with the lakeshore, which has a reliable pastry shop and the friendly wine bar, Da Giannino. If planning a picnic for the following day, call in at Stresa's sole supermarket, located on Via Roma, just off Piazza Cadorna. For dinner, try to avoid the lamentable restaurants on Piazza Cadorna in favour of a lively supper cruise or an elegant dinner at Hotel Regina Palace (*see Accommodation, page 92*).

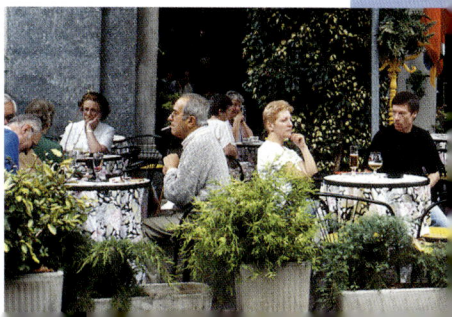

Above: Gothic frescoes in Santa Caterina
Right: relaxing at Caffe Nazionale

3. VILLA TARANTO & LAKE MERGOZZO *(see map, p22)*

From Stresa, take a short boat trip to the Villa Taranto in Pallanza-Verbania. Walk or cycle 6.5km (4 miles) through a nature reserve to Lake Mergozzo for a swim, a late lunch and stroll around the fishing village.

From the jetty on Piazza Marconi, take a morning ferry to Villa Taranto (45 minutes) or Pallanza (30 minutes and a short walk); the former is more direct but less frequent, so take the first boat in the right direction. To hire cycles, call the tourist office (0323-30416) the day before. If you want to take a taxi (or infrequent bus) from Pallanza to Mergozzo, visit Pallanza's tourist office by Villa Giulia, whose staff will advise you. For a beach picnic, buy provisions from the supermarket in Via Roma, Stresa. Alternatively, book lunch in La Quartina in Mergozzo (see Hotels, page 93). Return by the same route to Pallanza or take a bus or taxi.

The ferry might stop at **Baveno**, up the coast. Set in the lee of a pink granite mountain, Baveno is a small, subdued version of Stresa that has been popular with British visitors since Victorian times. If arriving at **Pallanza**, ask at the tourist office on Corso Zanitello about buses and taxis to Mergozzo. Next door is Villa Giulia, a frothy pink concoction built in 1847. Its gardens command views of the private islet of San Giovanni, which Toscanini acquired from the Borromean family.

Mussolini's Influence

Pallanza, the old part of town, was absorbed by Verbania in 1939, when Mussolini rechristened the resort Verbania as part of his campaign to revive the glories of ancient Rome. Pallanza was the only town on the lake not to fall under Borromean sway. Medieval Pallanza, with its smattering of bars and clubs, tends to attract elderly visitors in spring and younger ones in summer.

From Villa Giulia, follow the fashionable lakeside promenade round the promontory to **Villa Taranto** (9am–5pm), which has some of the finest gardens in the region. In 1931, Neil McEachern, a retired Scottish soldier, bought the property and devoted the rest of his life to planting and landscaping its gardens. Prompted by the mild climate, McEachern imported exotic plants: coffee, cotton and tea, lotus blossom, giant Amazonian water lilies and papyrus. Manicured tulip-lined flowerbeds contrast with exotic aquatic plants, a wooded ravine, and a soothing water garden's fountains and ponds.

The gardens have more than 20,000 plant species and are equally lovely in spring and autumn. In April and May cherry blossom floats over violets, narcissi and crocuses; camellias, azaleas, irises and rhododendrons all flourish.

Neil McEachern **(above)** devoted his life to Villa Taranto **(right)**

Summer is the time to see aquatic plants, oleanders, hydrangeas, roses and citrus fruits; in autumn you can see coppery Japanese maples, flowering shrubs, mellow dahlias and views of the terraces framed by russet leaves.

Since McEachern left the villa to the state, Villa Taranto has been a venue for political summits (held in the villa in the park, which is closed to the public), each marked by a tree-planting ceremony. Trees have been planted here by Margaret Thatcher, Helmut Kohl and Giulio Andreotti.

Retrace your steps to Pallanza and follow the scenic **Via Troubetskoy** west to the far side of the resort. From here you can watch the water sports around **Suna**. This former fishing village buzzes with beach life, though the occasional villa and Romanesque church hark back to a quieter era. Towards the end of Via Troubetskoy, the promenade merges into the coastal road, for the only tedious stretch of the walk, and Via 42 Martiri. Beyond a medieval tower and chapel on the right, take a footbridge across the Toce river to the **Fondotoce nature reserve** and follow the path through these protected wetlands to **Lake Mergozzo**, passing a war memorial just before the beach.

The shore is lined with reed beds but the **beach** itself is clear, peaceful and accessible. (Lake Mergozzo's beaches are secluded, and its water is cleaner and warmer than in other resorts.) The small lake is framed by forests, the surrounding hills are dotted with marble-carving workshops. If not picnicking on the beach, have a swim before dining in style on the lakeside terrace of La Quartina (tel: 0323-80118). Hugging the shore on Via Pallanza, this, the best hotel-restaurant on the lake, serves trout and perch from the lake, salami, cheeses and game from the hills, and a traditional Piedmontese tasting menu. This delightfully relaxing place overlooks a grassy beach, giving you another chance for a swim.

A Picturesque Hamlet

After lunch or a swim, follow Via Turati to the characterful village of **Mergozzo**, which lies a further 4km (2½ miles) west along the lake. The forested hills around the lake are quite atmospheric, even if one side is scarred by former quarries that used to produce the marble from which Milan cathedral was built. Café Portici, on the main piazza, is the place from which to see the picturesque hamlet that curves around the fishing harbour; from here, inviting alleys and flights of steps lead to a cloistered church and old-fashioned inns.

For dinner, you might opt for pasta with scampi or river shrimps at La Terrazze Due Palme (tel: 0323-80112) overlooking Lake Mergozzo. Alternatively, if not footsore, consider a supper back on Lake Maggiore: Boccon di Vino is a homely inn by the landing stage at Suna in Pallanza *(see Eating Out, page 76)*. If you are especially fit you could plan a demanding cycle ride from Villa Taranto to Cannobio along the lakeside path.

Right: Villa Taranto's exotic gardens contain over 20,000 plant species

4. LAKE ORTA *(see map, p22)*

A day trip to this romantic lake on the Orta San Giulio peninsula. Lunch at an inn before driving along the western shore of the lake to the sanctuary of Madonna del Sasso. Dine in the Moorish Villa Crespi (tel: 0332-911902; see *Eating Out, page 77*) in Orta before returning to Stresa.

Drive from Stresa to the eastern shore of Lake Orta, via the lakeside town of Omegna. Avoid Orta on Sundays, when it is congested by coach parties. If you don't have a car, consider booking an organised excursion from Stresa or Orta through the Orta tourist office (tel: 0322-905163).

For all its popularity, with Italians and foreigners alike, **Lake Orta** has managed to retain a certain mystique. A dreamy pocket of Piedmont, it includes the medieval village resort of Orta San Giulio and the enchanting Isola di San Giulio in the centre of the lake.

From Stresa, the route crosses the undulating hinterland between Lake Maggiore and Lake Orta, following the main road to Gravellona Toce. From there, take the ss229 to Omegna, a red-tiled town in the foothills of the Alps, and hug the eastern shore to Orta San Giulio. Dedicated shoppers might take a slight detour in **Omegna**, a small industrial town with a metal-working tradition dating back to the 18th century, when pewter was produced in the region. Pewter has since given way to brass, silver, aluminium and stainless steel, but household goods continue to be manufactured on these shores. Omegna is home to design guru Alessi, which is known for its stylish, funky kitchenware. The idiosyncratic products range from elegant, streamlined coffee pots to corkscrews cast in female form. Alessi's HQ and factory outlet are in the Crusinallo suburbs *(see Shopping, page 73)*. The Forum (tel: 0323-866141, closed Mon and Sun am) is a regional design showcase nearby.

Way of the Cross

From Omegna, the scenic lakeside road leads to the east bank's **Orta San Giulio**, a medieval village set snugly on a peninsula. The village has long been a fashionable if discreet resort known for its chic hotels as much as its soft light and air of spirituality. The approach road to Orta passes the entrance to **Sacro Monte** – leave the car in the higher of the two car parks, which is conveniently situated for access to the pedestrianised historic centre. *Sacri monti* (holy mounts) are a prominent feature of this corner of the lakes, and Orta's, the finest, is rivalled only by the one in Varese. Linked to tiny chapels, *sacri monti* are the climax of Franciscan devotional routes

that evoke the symbolic journey through the Holy Land. Set on a wooded hillside, Sacro Monte is a station on a *Via Crucis* (Way of the Cross) that wends through a series of 20 frescoed chapels that trace the history of St Francis. Slate-roofed Renaissance and baroque chapels are full of devotional paintings and statues by Lombard artists.

Left: a ferry arrives in Orta San Giulio

lake maggiore

The holy mount is well worth visiting for its peaceful atmosphere and misty views and the prospect of a light lunch in the inn just within the sanctuary. The Sacro Monte inn (tel: 0332-90220) is the place for ravioli, risotto and cold cuts; Taverna Antico Agnello (tel: 0322-90259, *see Eating Out, page 77*) is a rustic-style place in the heart of the village.

Spring Concert Festivals

From Sacro Monte to the village, take the steep Via dei Cappuccini path down to the shore, or the gentler but equally atmospheric Via Gemelli, which emerges outside the baroque Chiesa dell'Assunta – host of a spring concert festival complete with full historical pageantry. The church commands a dramatic view of the sloping **Salita della Motta**, which winds down to the main square, passing geranium-bedecked balconies and Renaissance palaces with peach-coloured facades.

The descent ends on **Piazza Motta** by the quaint waterfront, lined with outdoor cafés and old-fashioned hotels. As the hub of village life, the piazza is particularly animated during the Wednesday market. Competition is provided by Negri, a delicatessen that sells a variety of local produce, such as honey, salami, mushrooms and cheeses. Overlooking the square is the arcaded former town hall, the Palazzo della Comunità, whose frescoes feature a serpent symbol that relates to Orta's mythical origins. This facade is an exception to the traditional muted colours; buildings here tend to be painted in ochre, soft green or pink; white is forbidden. Take **Via Olina**, the main thoroughfare, if you want to indulge in some idle window-shopping in the medieval quarter. Not far away, on Largo de Gregori, you will find Rovera, a whimsical salami shop that features some-

Top: Franciscan statue at Sacro Monte, one of the stations on the Way of the Cross
Right: frescoes on the Palazzo della Comunità, Orta Giulio

what tasteless paintings of piglets tucking into a pork feast. The majority of visitors feel themselves drawn back to the waterfront where, bathed in soft light, the **Isola di San Giulio** beckons.

At the picturesque jetty, friendly boatmen guide ferries and motorboats to the island, which is a haven of tranquillity and supplication. In the convent at the heart of the isle, which is unfortunately closed to the public, blue-robed Benedictine nuns devote themselves to contemplation, work, prayer, and general shunning of the outside world. Chattering visitors clambering out beside the island **basilica** are soon hushed by the sombre mood of this Romanesque church. This seemingly enchanted island was supposedly over-run by serpents and dragons until AD 390 when Giulio, a Christian preacher, succeeded in banishing them before erecting a basilica in celebration.

Fragments of the the 5th-century church are visible in the crypt. The black pulpit, made of local serpentine marble, is a masterpiece of medieval sculpture. Inspired by saintly lore, the work depicts a seething mass of monsters, from grotesque griffons to drag-ons devouring one another's tails.

Spiritual Injunctions

The circular **Via Giulino** that hugs the high walls of the nunnery carries Buddhist-like injunctions to follow the right path. In a bid to evoke 'the island within', the abbess has created two pilgrimage paths – the Way of Meditation and the Way of Silence. 'Each journey begins with yourself' merges into 'Listen to the silence'; 'Listen to the water, the wind and your steps'; 'Silence is music and harmony'; and 'Silence is the language of love'.

Further along, opposite the entrance to the convent, the cosy San Giulio inn might be an appropriate place at which to ponder on 'the moment [that] is present, here and now'. You might enjoy a drink on the lakeside terrace. The island for-sakes its silence only in June, when a festival of ancient music is staged with bold historical pageantry.

If you have time, take the short drive to a sanctuary overlooking the lake. From Orta San Giulio, head south along the shore to Gozzano and follow the western shore to **San Maurizio**, an area known as 'tap country' after the presence of Italy's finest tap and bath makers. Based around the lake, manufacturers prospered on orders from Arab sheiks, but success exacted an ecological price. Lake Orta was polluted with industrial waste until the early 1990s, when a cleaning programme that coincided with a collapse of the gold tap market in certain sheikdoms resulted in the waters becoming fit for swimming once more. Resist the tap museum in favour of the sanctuary of **Madonna del Sasso**, which, just 2km (1 mile) away above Pella, and clearly signposted, forms the core of a pleasant series of hamlets. The sanctuary, built on a granite outcrop over the lake, is a frescoed baroque

Above: Basilica San Giulio has a sombre atmosphere

church with fine views over mountains and lake. You can return to Lake Maggiore by continuing along the western shore, passing Pella, Nonio and Omegna, before travelling inland via Armeno and Gignese, near the Monte Mottarone ski resort.

On the other hand, if you have a taste for the exotic, Orta San Giulio is where you will find the magical hotel-restaurant of **Villa Crespi**, a Moorish fantasy erected in 1879 by a local cotton merchant inspired by his travels to Baghdad and Persia. Villa Crespi is conveniently situated at the entrance to town, so parking is easy.

If by any chance you find that your senses are befuddled by the end of the evening, you might draw comfort from the admonition of the philosophical abbess on Isola San Giulio: 'When you are aware, the journey is over.'

5. LAKE VARESE AND LAKE LUGANO: VILLAS AND GARDENS AROUND VARESE *(see map, p36)*

A full-day country and lakeside drive from Varese, nudging the Swiss border on Lake Lugano. The tour begins with Villa Litta-Panza's contemporary art collection before a visit to Varese's nature reserve, then lunch in Luino on Lake Maggiore. The route then hugs Lake Lugano in an easterly direction before returning to Varese. Dine in the historic Villa Castiglioni just north of Varese, or on Lake Varese.

Begin at Villa Litta-Panza in Varese. From Piazza Monte Grappa in the centre of Varese, take Via Veratti/Via Argugiari north for about 1km (½ mile) and turn left along Via Castiglioni, following signs to Biumo Superiore and Villa Panza. Book a late lunch in Luino on Lake Maggiore and dinner in Villa Castiglioni, on the road back to Varese (see Accommodation, page 93). Check whether the Campo dei Fiori nature reserve (tel: 0332-235491) is open before you set out. The same applies to Villa Cicogna (tel: 0332-471134), but be warned that it becomes very busy on Sunday and in August. It's a good idea to bring your passport in case you find yourself seduced by the temptation to visit Switzerland.

Varese has long been overshadowed by the popularity of Stresa and Como. But the cleaning of Lake Varese,

Top: Villa Crespi hotel – a Moorish fantasy
Right: Villa Litta-Panza is full of modern art

lake maggiore

coupled with an effective promotion of its villas and gardens, is starting to attract discerning visitors. Known as a garden city, Varese has particularly peaceful lakes, from the eastern shore of Lake Maggiore to the scenic Italian stretch of Swissified Lake Lugano.

American Abstracts

Villa Litta-Panza (10am–6pm, closed Mon), located in Biumo Superiore – a leafy northern suburb of Varese – is a frescoed 18th-century mansion filled by a stunning array of 20th-century art. The owner-collector, Giuseppe Panza, was particularly keen on American abstracts, and the collection, with works from the 1950s onwards, is strong in work from the 1980s and 1990s. The best-known works, including abstracts by Sims and Simpson, and installations by Flavin, feature on a clear audio-guided tour.

Rather than disperse his eclectic collection among his five sons, Panza left it to the Italian equivalent of the National Trust. As both a listed monument and an art gallery, the villa appeals to both lovers of contemporary art and fans of classical architecture alike. The former will appreciate the abstract tonal canvases, rooms bathed in brash neon light, the bizarre installations, African masks and ceilings exposed to the elements. Classicists will admire the old Tuscan chests, the empire-style dining room, and

Lake Varese and Lake Lugano

a magnificent ballroom hung with chandeliers and subdued abstracts. The two aesthetics do come together at some points, as in the extended sightlines suggested by a long corridor that blends into a leafy bower in the garden. Beautifully set on the crest of a hill, the landscaped classical gardens provide a soothing break from the bold artworks. The coffee shop is your last chance to decide whether you belong to the contemporary-art camp or the classical-villa camp.

Nature Reserve and Observatory

If the weather allows, consider making a short visit to the **Campo dei Fiori** nature reserve and astronomical observatory. (From the villa, return to Viale Argugiari and head north along the same road for about 6km [4 miles], following the signs to Sacro Monte and Campo dei Fiori.) A winding road climbs to the summit, passing the devotional shrines of the **Sacro Monte** and affording vertiginous views down over **Lake Varese**, with its reeds and rushes. This is the heart of Varese's villa zone, and the hills are dotted with *belle époque* and Art Nouveau concoctions perched on grassy knolls.

The Campo dei Fiori reserve and science centre is a labour of love masterminded by Salvatore Furia, who personally defused or detonated thousands of unexploded mines to build the road to the reserve. His long-term ambition was to instil in young people his own passion for all aspects of the natural world, from the plants in the ground to the stars in the sky. To this end, the botanical gardens are an introduction to the surrounding landscape, with woods and valleys framed by the Alps. The observatory specialises in photographing comets through the lens of a powerful telescope.

Top: the stunning nave ceiling of the Sacro Monte
Right: Lake Lugano zigzags across the Swiss border

Next on the itinerary is Ponte Tresa on **Lake Lugano** (from Campo dei Fiori retrace your steps, taking the first road on the left that joins the ss233, the main road north to Ponte Tresa). Lake Lugano, which zigzags across the Swiss border, is wilder and less majestic than Lake Maggiore. Flat waters lap against daunting banks and the waterfront, with a steep shoreline, is often inaccessible. **Ponte Tresa** is a steamer stop and border village.

From Ponte Tresa head west to **Luino** on Lake Maggiore for lunch and a lakeside stroll or to visit the huge Wednesday market. Encircled by woods, Luino was once a centre for smugglers, whose contraband coffee and cigarettes would pass between Switzerland and Italy as market forces dictated. After lunch, retrace your steps to Ponte Tresa and follow the shore of Lake Lugano as far as **Porto Ceresio**, one of the most scenic drives in Varese province. A pretty Italian outpost, Porto Ceresio centres on an elegant harbour framed by steep shores. If time permits, consider a ferry crossing from the Italian port to its Swiss rival, the picturesque village of **Morcote**, set on the tip of the peninsula that runs down the lake towards Italy.

A Sforza's Gratitude

From Porto Ceresio, take the ss344 south to Varese, stopping at Bisuschio to visit **Villa Cicogna-Mozzoni** (open Sun and Aug 9.30am–noon, 2.30–7pm, or by arrangement), an impressive Renaissance stately home now inhabited by the genial Count Cicogna-Mozzoni. The family fortunes were founded on a happy accident: in 1476, Galeazzo Sforza, the powerful Duke of Milan, was out hunting when he chanced upon a bear on the rampage. He was saved from certain death by Agostino Mozzoni and his dog. To express his gratitude, the Duke of Milan funded a scheme to transform Mozzoni's simple hunting lodge into this lovely villa, complete with Italianate gardens, formal box hedges and fountains.

When contemporary funds are made available, the grand water stairway should be a romantic cascade once more, and the restoration of the frescoes and fountains complete. In the meantime, the bohemian, jazz-loving Count Cicogna-Mozzoni continues to host receptions as a way to finance the return of his home to its full Renaissance splendour. To end the day in style, dine in Villa Castiglioni (tel: 0332-200201), a sumptuous 18th-century villa-hotel at Induno Olona, just north of Varese on the road back from Villa Cicogna.

Top: the picturesque Swiss village of Morcote
Right: catching up on news in Varese

Lake *Como*

6. VILLA CARLOTTA AND BELLAGIO *(see map, p40)*

A morning expedition to Villa Carlotta's gorgeous gardens at Tremezzo, followed by a short ferry ride to Bellagio, the most atmospheric resort in the lakes. Enjoy lunch with a view before a stroll around the gardens of Villa Melzi and a swim or some shopping.

By boat from Como to Villa Carlotta (Tremezzo or Villa Carlotta stops; allow 90 minutes by ferry or 30 minutes by hydrofoil), then onto Bellagio; return to Como by hydrofoil; the last one leaves at about 8pm.

Historically adored by the British, **Lake Como** is now more popular with Americans, due to its superb hotels, clever marketing, and the mystique surrounding Bellagio. (As a tribute to this tiny lakeside resort, Las Vegas has its own, splendidly kitsch, Bellagio Hotel.) Picturesque Bellagio is Lake Como's calling card, with Villa Carlotta and its glorious baroque gardens the highlight. This central stretch of the lake, embracing Menaggio, Bellagio, Tremezzo and Varenna, is definitely the most seductive.

Seducing Marilyn Monroe

Henry James was well aware of Lake Como's reputation for illicit trysts: 'It is commonly the spot to which inflamed young gentlemen invite the wives of other gentlemen to fly with them and ignore the restrictions of public opinion.' It was here, according to locals, that President Kennedy romanced Marilyn Monroe.

The resort of **Tremezzo** had its heyday in the 1930s but it retains a courtly grace, epitomised by the Grand Hotel Tremezzo. This delightfully mothballed affair remains popular for wicked weekends and whirlwind tours for gardening clubs alike. This part of the lake is known for its azaleas, camellias, rhododendrons and hydrangeas, all best appreciated during the spring.

Villa Carlotta (9am–6pm), a short walk from the Tremezzo jetty, was a wedding present from a Prussian princess to her daughter, Princess Carlotta of Nassau. Carlotta established a small court here and completed the landscaping of the gardens in the 1860s. As a result, guests are greeted by a prosperous baroque villa bordered by a profusion of pink and white azaleas and a theatrical staircase leading up to orange and lemon terraces. The focus is on the interplay between villa and lake, and the changing perspectives provided by the terraced gardens. The dazzling azaleas and rhododendrons

Right: heading towards Tremezzo on the local ferry

provide a colourful contrast to the villa's cool neoclassical interior. The freshly refurbished house inevitably plays second fiddle to the gardens, despite bursting with neoclassical statuary, including Canova's *Cupid and Psyche*.

The ornamental pool in front of the villa marks the way from whimsical grottoes to rose arbours, and to the rockery and palm collection on the banks above. Further up are clusters of heathers, magnolia and azaleas, a rhododendron grove and a moody glade of ferns, complete with rushing stream. The plants evoke an exotic map of the world: cedars of Lebanon, Egyptian papyrus, Japanese maples and banana trees, Chinese bamboo, Indian tea, Mediterranean agaves, ferns from New Zealand, eucalyptus and succulents from Australia, and giant sequoias from South America.

Bellagio Villa

From Tremezzo, frequent steamers cross the lake to **Bellagio**. This resort is set on the tip of Punto Spartivento, at which point the lake splits into two. On the bustling waterfront, peaches-and-cream houses lead to a promenade lined with oleanders and limes. There are red-roofed houses, pastel-tinged facades, steep cobbled alleys, quaint craft shops, a Romanesque bell tower

Top Right: Villa Carlotta's theatrical staircase leads up to citrus terraces
Right: the folly at the entrance to Villa Melzi

and lakeside vistas. Beneath this picture-postcard exterior lies an anglicised gentility leavened by the southern warmth of locals who love their village.

A gentle lakeside promenade leads to Villa Melzi, whose secluded sunbathing spots afford a solitude rare in Bellagio. Stroll past the **Lido** (10am–7pm), unless you want to stop there for a quick drink, or a swim, either in a tiny pool or in the lake. The lake's central part, around Bellagio, Menaggio and Tremezzo, is fine for swimming, with wind-surfing better at the top end. Before taking to the water, gather your thoughts in the most peaceful gardens in the lakes.

Villa Melzi (9am–6pm) is an austere neo-classical villa set in the first 'English' gardens on the lake. Though the villa, which is owned by Duke Gallirate Scotti, is a private residence, the romantic grounds enchant visitors. An atmospheric grotto opens onto Japanese water gardens cleverly concealed from the lake. An intimate mood is created by an ornamental pool framed by cedars, maples, camphor and myrrh. On the formal terraces above, classical statuary gives way to gently rolling lawns bordered by a pine grove. Below, standing guard by the lake, is a quaint coffee house that captivated Stendhal and Liszt. Beside this bold folly, an avenue of plane trees leads along the shore to the villa, chapel and boatyards, with lofty vistas interspersed with banks of camellias. Compared with Villa Carlotta, which is awash with colour, Villa Melzi favours subtle shades of green and a low-key mood, more in keeping with contemporary tastes.

Stroll back to the Bellagio waterfront and climb Salita Serbelloni to the church of **San Giacomo** at the top of the village. A Romanesque bell tower and curious altar notwithstanding, the church is disappointing. But you should see the craft shops and fashion outlets. Bellagioseta (Via Roma 13) sells hand-finished silk ties, some with copies of old masters, made in Como; Luigi Tacchi (Via Garibaldi 22) specialises in hand-carved bowls, boxes and games made from olive wood. Further on, Laved (No 41 and 60) makes

glass ornaments, including kitsch Christmas decorations. On the way back to the waterfront, you'll find Dolce & Gabbana, Fendi and Moschino goods at Principessa (Salita Serbelloni 16).

For **lunch or dinner**, avoid the waterfront tourist traps in favour of the best fish restaurant, Silvio (031-950322) on Via Carcano, or La Pergola (tel: 031-950263; book for dinner and weekends), facing the port in Pescallo. If you fancy an extravagant dinner on a luxury hotel's lakeside terrace, before a water-taxi ride home to reality, consider a trip to Villa Serbelloni *(see page 94)*. Otherwise, wait for the ferry in Bar San Remo on the waterfront.

7. COMO TOWN AND BRUNATE CABLE-CAR *(see map below)*

Explore the town of Como on foot, from the waterfront by Piazza Cavour to the cathedral. See glorious Gothic architecture, hit the shops for fashion and silk items, and take a cable-car ride to Brunate.

If you don't mind the considerable expense, make advance reservations for dinner in the Renaissance Villa d'Este in Cernobbio, the most prestigious villa hotel in the country (see Accommodation, page 94).

The biggest resort on the lake, **Como** represents a somewhat disconcerting combination of fascinating historic city, bustling commercial centre and hectic tourist resort. On the one hand Como has a wonderful medieval quarter and, conversely, some of the most lustrous shops of any resort in northern Italy. On the other hand there is nothing approaching the intimacy and timelessness of the small lakeside villages. Whether or not Como appeals to your idea of the perfect holiday resort, it is a convenient springboard for explorations of the lakes region.

Roman Origins

Since its earliest days, when it rose to prominence as a Roman town, Como has been an industrious, aspirational kind of place. The Romanesque style, which is

Above: sunbathing beside the lake at Bellagio's lido

particularly pronounced in Como, has been woven into the city's architectural fabric, from churches to fortified medieval towers. Many monuments also owe much to the craftsmanship of the *Maestri Comacini*, the medieval master-builders and sculptors who perfected the Lombard style.

The citizens of Como, who are known as *Comaschi*, seem to have couturier skills in their blood: they have manufactured silks, velvets, brocades and damasks since the 16th century. Although the silkworms are no longer bred around the lake, Chinese thread is woven and dyed here to the exacting specifications of the leading Milanese fashion houses. Mantero and Ratti, two of the city's great silk families, supply big-name designers such as Chanel, Dior, Versace and Yves St Laurent. While the superstar fashion houses virtually all have showrooms in Milan, many either come from the lakes or – like the Versace dynasty – have villas on Lake Como. The silk and textile industry has made an immeasurable contribution to Como's prosperity and gracious living.

The Old Centre

Piazza Cavour, the city's waterfront square, is the undoubted focal point for tourists. Flanked by the terraces of outdoor cafés, it is always bustling with ferry traffic. A little way inland, **Piazza del Duomo** represents the best introduction to the medieval quarter, centred as it is on the splendidly solemn cathedral. Work on the **Duomo** began in 1396, and it has recently been restored. In particular, look out for its impressive gabled facade, and observe how it spans the transition in styles from late Gothic to Renaissance, with a richly sculpted main portal. Devoted Catholics might be interested to know that Mass is celebrated in the cathedral every hour. Flanking the cathedral is the bell tower and the magnificent **Broletto**. Formerly the town hall, this is an elegantly arcaded Gothic affair complete with triple-arched windows.

From here, take **Via Vittorio Emanuele**, which is the main shopping thoroughfare, to **Piazza Medaglie d'Oro**, stopping at selected silk shops and emporiums that feature the latest fashions. Beyond the world of silk, Como's temples of consumerism might well tempt you with their high-class jewellery, shoes and wonderful luggage and handbags, as well as the newest designs from Milanese studios. One recommendation is Picci, at No 54, which supplies silk for Armani. This fine outlet has been here for four generations, and produces crepe de Chine scarves as well as hand-finished silk ties. Other classic silk shops are on Piazza San Fedele and Piazza Cavour. Bear in mind that the greater the number of colours used within the design, the higher the cost is likely to be.

From Piazza Medaglie d'Oro, head back towards the waterfront. If you want to take a break for a coffee at this point, check out Aida, a historic

Right: the richly decorated right transept of Como's Duomo

café on Piazza Fedele. Facing the square stands the remodelled Romanesque church of **San Fedele**, which was, in all probability, the Como's first cathedral; it is worth walking to the rear of the square to admire the striking apse. Continuing towards the lake, stroll down **Via Luini**, which has another excellent ice-cream parlour: indeed Bolla is considered to be one of the best in Como.

Como's Most Famous Son

A little further along, turn left into **Piazza Volta**, which is flanked by boisterous outdoor cafés. The square has a statue of Count Alessandro Volta (1745–1827), a self-taught physicist who invented the battery and the units of electricity known as the volt. In summer the square serves as a venue for concerts and occasional sightings of George Clooney, Como's favourite son.

If the weather is fine, you might consider taking lunch in the elevated environs of Brunate, which is set high above the city. Follow the lakeshore north to **Piazza de Gasperi**, passing the city's most upmarket cafés and hotels. The cable car to **Brunate** leaves at half-hour intervals, and the journey to the summit takes seven minutes. Once at the top, you can enjoy a rustic lunch at Ristorante Falchetto on Salita Peltrera. From Brunate there are outstanding views of the Alps, the lake and an array of 19th-century villas. This is also a good vantage point from which to see the city's octagonal Roman layout.

On a summer's evening, join the throng on a stroll along the waterfront from **Piazza Cavour** to the stately **Villa Olmo**, which is set in splendid Italianate gardens. For dinner, choose between the splendour of the lakeside Villa d'Este in Cernobbio, and the atmospheric Le Due Corti (tel: 031-328111), a converted convent on Piazza Vittoria, in the heart of town.

Above: taking life easy on Via Vittorio Emanuele
Left: just browsing

8. ISOLA COMACINA AND VILLA BALBIANELLO *(see map, p40)*

A romantic cruise, lunch on a haunted island, and a short boat trip to see Villa Balbianello and the finest setting on the lake.

Catch a ferry from Como to Sala Comacina, where a waiting boat will make the two-minute crossing to Isola Comacina. After lunch, ask a boatman to make the short trip to Villa Balbianello (tel: 0344-56110; gardens Thur–Sun, Tues 10am–1pm, 2–6pm; villa by appointment only). Book a visit to the villa in advance, and maybe reserve a table at the Locanda dell'Isola for lunch (tel: 0344-55083, open Mar–Oct, closed Tues in spring).

Lake Como is a haven for celebrities, who tend to be discreet, and whose privacy is respected. Frequent visitors, often linked to the Versace fashion dynasty, include Elton John, Madonna and Bruce Springsteen. George Clooney is the latest celebrity to have ensconced himself in a lovely lakeside villa – close to Villa Versace. Villa Balbianello featured as a setting for the film *A Month on the Lake*, starring Vanessa Redgrave and Uma Thurman, and it also appeared in *The Phantom Menace*, the latest *Star Wars* epic.

The **steamer cruise** leads to the central part of the lake, famed for its villas and gardens. En route, it criss-crosses the most beautiful arm of Lake Como, moving from the sunny western side to the shady eastern side. The west has the best resorts, historic villas and gardens; the wilder east is dotted with Romanesque churches and the odd silk factory. **Cernobbio**, on the western shore, is the first major resort after Como. The town has an appealing lakefront and old quarter, but the most eye-catching sights are the Renaissance **Villa d'Este**, the most prestigious hotel in the lakes, and the 19th-century **Villa Erba**, which belonged to the family of film director Luchino Visconti before becoming a conference centre. **Moltrasio**, also set on the sunny side of the lake, features the deceptively low-key **Villa Versace**, designed as a perfect setting for fashion shows. Gianni Versace is buried here.

The 'Amalfi Coast'

The 'Amalfi coast', so named because its dramatic cliffs and wild atmosphere are reminiscent of the real Amalfi coast near Naples, begins just to the north, on the eastern shore. The fine Romanesque church in **Pognana** is shaded by the villas on the sunny side of the lake. **Nesso**, also on the shaded side, is known for a long Romanesque bridge and a gloomy five-storey silk factory that closed in the 1950s, when silk production moved to China, Brazil and Turkey. From Nesso, the ferry crosses to **Argegno**, with its redtile roofs and sunny lakefront. Perch abound in this, the lake's deepest section.

At **Sala Comacina**, transfer to a small boat for a ride to the tiny island facing the port. Here you will find the Locanda

Right: the Villa d'Este, the lakes' leading hotel

dell'Isola restaurant, which gets very busy at weekends – aim to be there at 1pm for lunch, 8pm for dinner. The price of a set meal with wine includes the return boat ride from Sala. If you are going to be early for lunch, have a drink at Sala's lakeside bar before setting out. The Festival of St John, the most magical festival in the lakes, is celebrated on these waters on 24 June: as night falls, thousands of bobbing boats are illuminated by candles and fireworks. On the far shore, beyond the island, is the blighted fishing village of **Lezzeno**, in contrast to splendid **Bellagio** which commands the point beyond. Due to the gigantic shadow cast by the mountains, the sun is permanently blotted out, and the locals pay the price in terms of tourism.

The 'haunted' island of **Isola Comacina** just off Sala is the place for lunch in a rustic inn. The island was cursed by the Bishop of Como in 1169, probably in revenge for its alliance with Milan, and Como's soldiers razed its churches and massacred the population. More recently the island belonged to Belgium's King Albert I. Since 1927 it has been Italian, but governed jointly. Hence the presence of Belgian and Italian flags.

Rite of Fire

In 1949 a passing English journalist by the name of Frances Dale apparently came up with the idea of a rite-of-fire exorcism to stimulate the tourist trade – and the word spread like wildfire. Today guests are welcomed by the current, melancholic, master of ceremonies, the bobble-hatted Benvenuto Puricelli (who was born in Sala Comacina, facing the island, and who served a stint as head chef at London's Penthouse Club). Puricelli has been exorcising the island's demons on a daily basis for the past 25 years, in which time the roster of visiting international celebrities has included Sylvester Stallone, Michael Schumacher, Arnold Schwarzenegger and Ruud Gullit. The inn's gallery is full of portraits of such stars participating in the theatrical 'purification ceremony'. For all the hokum, the set meal – ranging from an array of *antipasti* to chunks of parmesan, baked onions, grilled trout, chicken, and orange ice cream – is very good. The rite of fire begins at the flambéed-coffee stage and involves arcane incantations and copious amounts of brandy and sugar. Even if the ceremony is not your cup of *capuchin*, the lakeside views are simply gorgeous. After lunch you might want to follow a path that passes the ruins of the Romanesque churches sacked in 1169. A boatman at Isola Comacina will ferry you to the next attraction on the agenda.

Above: Nesso, on the shaded side of the lake
Right: Benvenuto Puricelli exorcises demons

Situated on a promontory between Lenno and Sala Comacina, **Isola Balbianello** enjoys the loveliest setting on the Italian lakes. Romance, peace, seclusion, tranquillity, fabulous gardens: this ochre-coloured villa has everything you might hope to find in the lakes region. The last resident was the wealthy explorer, Count Guido Monzino, who left the villa to the FAI, the Italian equivalent of the National Trust in 1988. The villa was originally a Franciscan foundation, the venue exemplifying the friars' predilection for remote locations on lakeside promontories.

A Cardinal's Retreat

The lowest section of the three-tiered structure incorporates the medieval Franciscan church, including a quaint pair of bell towers. In 1786, the villa was bought by the sybaritic Cardinal Durini as a retreat from his taxing diplomatic missions. After landscaping this rocky spur, he enlarged the villa and, as his crowning achievement, added the loggia on the highest point. Durini wanted both a memorable venue for literary salons, and a place from which to admire the sun setting in the mountains. Count Monzino was also entranced by the loggia: using it as a study, he filled it with Polar maps, Himalayan prints and mountaineering memorabilia.

The vine-clad villa is dotted with souvenirs of the count's adventures, including the sledge on which he trudged to the North Pole in 1971. Fine though the interior is, the loggia is more impressive when seen from the outside: it is covered in climbing plants, and an old fig tree clings to its columns. Oppressive empire-style aesthetics pervade the smoking room, master bedroom and library. In contrast with the gloomy bachelor quarters, Monzino's mother's rooms are cheerful and open to the lake.

The **garden** makes the most of its setting on a rocky spur, and the poor soil conditions. Somewhere between a classical 18th-century Italian affair and a romantic English garden, it exploits the gnarled outline of the barren rock to present beguiling paths that lead up to a three-arched folly. Bound by boxwood and laurel hedges, pergolas, climbing plants and scented wisteria, this secret garden is mirrored in the lake. Cypresses, holm oaks and plane trees pruned into candelabra shapes provide shelter for snowdrops, cyclamen and magnolia. Closer to the villa are lakeside vistas framed by terracotta tubs of pink hydrangeas and classical statues that protrude from beds of azaleas.

The final view is of the Italian flag unfurled on the jetty, in keeping with Monzino's will. The explorer wanted the gesture to be 'in memory of all the flags my alpine guides placed on countless peaks all over the world'. Unless you are up for more adventures, ask the boatman to drop you off at the **Lenno** jetty, where you can pick up the ferry to your accommodation.

Right: a ferry operator waits for passengers at Sala Comacina

9. BERGAMO *(see map below)*

Visit an art gallery in the lower town before exploring the upper town.

The upper town is pedestrianised, and parking in Bergamo is tricky, especially on Sunday, so travel by train (direct from Milan or Como). From the station, catch bus 9a, 9b or 9c to the Accademia Carrara in the lower town.

Seen from afar, **Bergamo** is a mass of bell towers and domes silhouetted against the snow-capped Alps. A well-preserved medieval town, Bergamo is admired for its cool Venetian beauty and intimate squares. Originally settled by Celts and Romans, the town nevertheless has a Venetian soul: some 400 years of Venetian rule have left their mark in the graceful architecture and symbols of Venice, even if the harsh alpine setting and complete absence of a lagoon are distinctly non-Venetian. There are two distinct centres: the beguiling *città alta* (upper town) and the more modern *città bassa* (lower town). The former is bound by a circle of 16th-century walls built by the Venetians after the city expanded beyond its medieval ramparts and fortress.

Birthplace of Trussardi

This itinerary begins in the **lower town** with a visit to the **Accademia Carrara** (10am–1pm, 3–6.45pm, closed Mon; tel: 035-399643), which has a fine collection of Lombard and Venetian art amassed by a local aristocrat in the 18th century. Among the treasures in this neoclassical palace are Gothic works by Pisanello, and Renaissance and Mannerist works by Bellini, Veronese, Tiepolo, Tintoretto, Raphael and Mantegna. Apart from the gallery, only shopping and the cafés are likely to detain you in the lower town. The birthplace of the Trussardi fashion dynasty, Bergamo abounds in designer shops. The most elegant shopping district is the arcaded **Il Sentierone**, a pedestrianised promenade lined with cafés and fashion outlets.

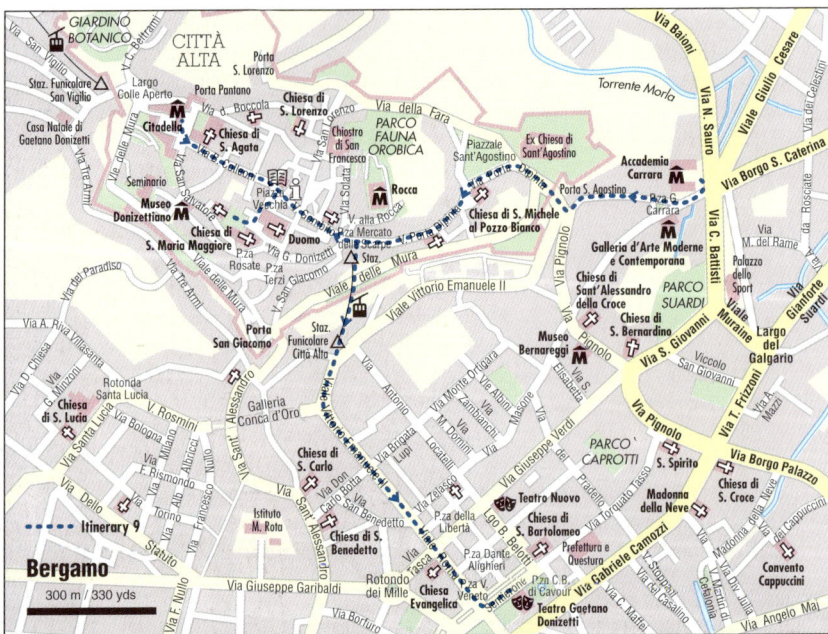

lake como

From the gallery, follow Via della Noca through the Venetian gateway of Porta Sant'Agostino into the **upper town** and the **Piazza Vecchia**, which is often described as the most perfect square in Italy. En route, there are views of the Venetian ramparts and the cable-car station. The adjoining squares of Piazza del Duomo and Piazza Vecchia present a harmonious Gothic ensemble. Piazza Vecchia in particular, is a charming composition of cobblestones, red-tiled rooftops, a tiny fountain and slender clock tower.

The Religious Heart

Flanking the porticoed square is the **Palazzo della Ragione**, the much remodelled medieval council chambers, decorated with a winged Lion of St Mark, the symbol of Venice. The **Piazza del Duomo**, the religious heart of town, is home to the Gothic **cathedral** and, submerged beneath a sumptuous baroque interior, the Romanesque basilica of **Santa Maria Maggiore** (9am–noon, 3–6pm). Attached to the south wall is the ornate **Cappella Colleoni** (9am–noon, 2.30–6.30pm). Built in honour of a wealthy mercenary, this Lombard Renaissance chapel is a jewel box of a mausoleum.

Bergamo is known for its earthy cuisine: Agnello d'Oro (Via Gombito 22, tel: 035-249883) is one of the most atmospheric eateries in the upper town. Set in a 17th-century palace, it serves filling regional dishes such as *casoncelli* (ravioli) and polenta with *taleggio* and salami. Wander along Via Colleoni to the **Citadella**, a former residence of Venetian captains and sailors, past alleys lined by Renaissance and baroque palaces.

To return to the lower town, the **cable car** on Piazza Mercato delle Scarpe drops to Viale V Emanuele, whence it is a pleasant stroll past the shops and cafés on Piazza Matteotti and the adjoining **Il Sentierone**.

Above: Palazzo della Ragione, Bergamo
Right: Santa Maria Maggiore, Bergamo

Lake
Iseo

10. ISEO, MONTE ISOLA
AND THE LAKESHORE *(see map, p52)*

Explore the town of Iseo before crossing to the pretty island for a stroll and lunch, followed by a drive to Lake Iseo, with its wild western shore.

Check ferry timetables and cruise details at the tourist office (Lungolago Marconi, tel: 030-980209). The journey by car outlined below is equally lovely if taken on a cruise that stops at ports such as Lovere and Pisogne.

Unfairly neglected in favour of the larger lakes, **Lake Iseo** is more tranquil and less self-consciously quaint. Lake Iseo is short on the beautiful attractions that are a hallmark of the other lakes, but compensates with fine walks and a gentler way of life. **Iseo**, the best base from which to explore the lake, is a lovely historic town that hasn't completely sold out to tourism. Until the 1870s, it was a significant port that shipped grain from Valle Camonica and steel from the industrial lakeside ports. Today it is a commercial town on a smaller scale. Sandwiched between the waterfront and a feudal castle, it retains its cosy medieval street pattern and an elegant promenade.

The liveliest part of town, the porticoed **Piazza Garibaldi** is dominated by a statue of the great patriot perched on a mossy rock – this is one of the few horseless statues of the Risorgimento leader. Also on the square is Caffé Eden, an atmospheric bar and wine-tasting centre set in a converted cinema. For fine wines, its only rival is Enoteca Damiani on Piazza Mazzini, the next flower-bedecked square to the west. From here, Via Sombrico leads to the church of **Sant'Andrea**, which, although clumsily remodelled, has the finest Romanesque bell tower in the region. Further north, the recently restored **Castello Oldofredi** is set on a bland mound.

Returning to the waterfront via Piazza Garibaldi, you can sail to Iseo's peaceful island or to go window-shopping in an upmarket shopping centre based in a former textile factory. Shaded by plane trees, the promenade is a pleasant spot from which to watch the lazy lake traffic go by.

From Iseo (or Sulzano) take the relaxing ferry ride to Peschiera Maraglio on **Monte Isola**, which is usually

Above: the elegant lakeside town of Iseo
Right: 19th-century national hero Giuseppe Garibaldi

the second or third stop. The mountainous, densely forested Monte Isola, the largest lake island in Europe, supports a 1,700-strong community of fishermen, boat-builders and net-makers. The pace of life is palpably slow: private cars are banned – a minibus service connects the various hamlets. The fishing hamlets reveal refined touches, from sculpted portals to tiny court-yards and loggias. Above, tiers of olive groves merge into vineyards and chestnut groves. The highest peak is surmounted by a 16th-century sanctuary, built over a pagan shrine.

The best gentle walk is from **Peschiera Maraglio** to Sensole, both hamlets set on the sunny side of the island. The view encompasses the tiny island of San Paolo, with the town of Iseo melting into the background. From the landing stage, stroll along the lakeshore towards **Sensole** for an outdoor lunch in a typical lakeside haunt such as Vittoria (tel: 030-9886222, closed Thur) just before Sensole. En route, check out a cluttered shop that sells fishing nets and hammocks: net-making is an integral part of life on the island. An industry initiated by Cluniac monks 1,000 years ago now embraces Wimbledon tennis nets and World Cup football nets. Depending on the season and time of day, there may be signs of boat-builders at work or fishermen laying their catch out to dry in the sun. After you've had your fill of grilled sardines, perch risotto and lake scampi, wander along to the Sen-sole jetty, taking in views of com-petent swimmers competing for attention with struggling ducklings.

A Gourmet's Paradise

After the ferry ride back to **Iseo**, take a scenic drive clockwise round the lake to see Riva di Solto's rugged western shore in Bergamo province. By contrast, the Brescian bank on the eastern shore is more mundane. **Clusane**, just to the west of Iseo, is a food-lover's paradise: there are fine fish restaurants on the waterfront, and the Monday market in the main square features stalls

laden with cheeses and salami. Crowned by a castle, Clusane overlooks a busy port full of traditional red- or yellow-rimmed fishing boats setting out in search of tench, pike, chub and lake sardines. The village borders wine-growing Franciacorta *(see Itinerary 11, page 53)* and its hinterland is dotted with rustic inns, so all in all it is a gateway to gastronomic adventures.

Sarnico, the first resort on the west shore, occupies the site of a prehistoric stilt village and owes much of its character to the ruined medieval ramparts and graceful loggias. But it is best-known for the speedboat companies that support one of the lake's premier activities. In three generations, Riva (the Ferrari of speedboats), founded here, has moved from the construction of rowing boats to the manufacture of world-class speedboats and yachts.

From over-quarried **Tavernola** to **Riva di Solto** lies the most dramatic

Right: the serene and modest Lake Iseo

stretch of the west shore, with coves carved into limestone cliffs and sheer ravines running down to gnarled rocks. These jagged formations reputedly inspired Leonardo da Vinci's *Virgin of the Rocks* and possibly the *Mona Lisa*. Riva di Solto is a pretty fishing hamlet full of alleys and arches, and with placid views across to the domesticated shore of Monte Isola.

Lake Iseo

5 km / 3 miles

- - - Itinerary 10
- - - Itinerary 11

11. the franciacorta wine trail

Lovere, with its medieval ramparts, dominates the northern end of the lake. Lovere was a Venetian textile town before turning to steel, then tourism based on water-sports. Look out for its restored medieval towers and the frescoed Renaissance church of **Santa Maria** in **Valvendra**.

The medieval quarter of the former arms-manufacturing town of **Pisogne** exudes an air of benign neglect. Just north of the centre, the church of **Santa Maria della Neve** (9am–6pm, closed Mon; key from the adjoining Bar Romanino; tel: 0364-87032) is known as a poor man's Sistine Chapel. Its frescoes are by Romanino (*circa* 1484–1559), a Brescian artist known for his realistic portraits – which can be found all over Lake Iseo. The church's frescoes depict a typical Romanino scene of plump peasants.

For dinner, shoot past **Sulzano**, unless you intend to take the ferry back to Monte Isola for another fish feast. Many of the best restaurants are on the island, or in Clusane and the surrounding hills (*see Eating Out, page 78*).

11. THE FRANCIACORTA WINE TRAIL (see map, p52)

A leisurely drive from Lake Iseo through the wine-growing region of Franciacorta, stopping at a Cluniac monastery and vineyards.

Contact the local wine growers' association (tel: 030-7760870; www.strada delfranciacorta.it) to check which wine estates are open. As well as Villa (see below) try and see Bellavista estate (tel: 030-7762000) or Longhi de Carli (030-77602080). For lunch at L'Albereta (tel: 030-7760550), book in advance. On Sunday you can visit the Castello di Bornato medieval fortress/wine estate.

The prestigious wine-growing region of **Franciacorta** is to the south of Lake Iseo. This part of the country is known for producing sparkling champagne-style wines, dry, velvety whites and medium-bodied reds. The rolling countryside is dotted with fortified manor houses and elegant villas, many of which have been transformed into flourishing wine estates or inns. Franciacorta owes its prosperity to the work of those medieval monks who colonised this hitherto untamed corner of Lombardy.

In the 11th century, local nobles called on Cluniac monks to drain the land around Franciacorta. The results were beneficial to both parties: the local economy received a welcome boost, and the way was paved,

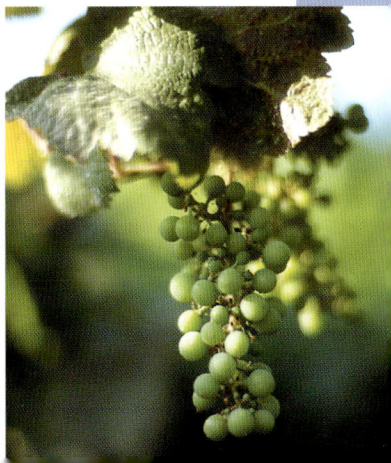

Above: the resort of Sarnico on Lake Iseo
Right: raw material for velvety white wines

literally, for the building of impressive new monasteries. The ecclesiastical authorities appreciated the local climate and countryside to the extent that the village of Borgonato became the summer residence of monks from Brescia's Santa Giulia monastery. As for the secular authorities, they granted tax concessions, which led to the region's contemporary nickname of Corte Franca (Free Court). These advantageous financial considerations encouraged patrician families from Brescia and beyond to build villas in the mellow hills here. Wealthy restaurateurs and viniculturists followed in their wake, and Franciacorta was on the way to becoming the sought-after retreat it is today.

You will find the Cluniac monastery of **San Pietro** (9am–noon, 3–6pm) and the peat bogs of **Le Torbiere** just outside **Iseo**. (From the centre of Iseo take the ss510 in the direction of Brescia, and the monastery of San Pietro appears on the right after about 2km/1 mile). The monastery, situated beside Lake Iseo and enveloped by marshes bordered by poppies and elms, has an appropriately other-worldly atmosphere. The Romanesque bell tower opens onto a sober interior relieved by Gothic frescoes and cloisters.

Le Torbiere Nature Reserve

Le Torbiere Nature Reserve (always open) at the foot of the monastery is the watery domain of perch, trout and eel. It is also home to a variety of predatory birds, from herons to kingfishers. You can stroll along the paths that wend their way through the peat bogs, perhaps catching glimpses of white swans gliding between the water-lilies, or the occasional swoop of a marsh falcon.

From San Pietro, turn right and follow the road south for 6km (3½ miles) to Bornato, to see the **Castello di Bornato** (Sun 10am–noon, 2.30–6pm; tel: 030-725006), a crenellated medieval castle that opens onto a Renaissance villa and Italianate gardens. The castle's small wine estate serves as a reminder that this is the beginning of a

Above: vineyards abound throughout the Franciacorta region
Left: a resident of the Cluniac monastery of San Pietro

wine route that progresses through the gently undulating countryside.

Follow the marked **Franciacorta wine route** east to **Monticelli Brusati** and vineyards that stretch out for as far as the eye can see. This is the delightful setting for the **Azienda Villa wine estate**, complete with wine tours, farm-stay homes and a welcoming rustic inn (tel: 030-652329; www.villa-franciacorta.it) that serves pasta, farm-fresh *antipasti* and superb wines.

From the estate, retrace your steps to Bornato and follow the meandering wine route to **Erbusco**, the unassuming stone-built village at the centre of the wine district. If you are not planning to dine in style at the Michelin-starred L'Albereta *(see Eating Out, page 78)*, you might want to visit the Cantine di Franciacorta wine shop or taste wines at the prestigious **Bellavista** or **Longhi de Carli** estates. If these are closed, try the attractively situated **Conti Castaldi** wine estate in neighbouring Adro then follow the wine route to **Borgonato** near Corte Franca to see Berlucchi Fratelli, the prettiest estate.

A return to the lakeshore at **Sarnico** marks the end of the wine route, but introduces the opportunity to partake in gastronomic dinners in **Clusane**, a fishing village on the way back to Iseo. Clusane is known for its baked-tench cuisine *(see Eating Out, page 78)* – the town has almost as many restaurants as fishermen. If there is time, you might want to take a swim at Sassabanek just before Iseo, where there are well-managed pools, a picnic area, and a grassy beach on the lake. You can easily return to Clusane for dinner.

12. VALLE CAMONICA'S PREHISTORIC ROCK CARVINGS *(see map, p56)*

From Lake Iseo, drive north to see the Valle Camonica rock carvings. Visit a spa resort and a prehistoric theme park before a late, rural lunch.

Book lunch in Il Riccio (tel: 0364-622113 or 0338-6083917) near Capo di Ponte, and consider booking a guide for the rock carvings (tel: 0364-42212/513609). If you wish to make this a full-day trip, you can book ahead for a visit to the White War Museum (tel: 0364-906420) in Temu.

From Lovere or Pisogne, leave **Lake Iseo** on a northerly drive to **Valle Camonica** to see the greatest concentration of prehistoric rock carvings in Europe. The valley route (along the ss42) passes a spa resort and medieval villages that have worked stone, wood and iron ore since ancient times. Rural and industrial at the same time, this distinctive valley has been inhabited since the Neolithic era, when the Camuni tribal civilisation first etched itself into immortality. These primitive carvings, known as 'stick men' by the valley's modern inhabitants, have always lurked in the local consciousness, though they were formally discovered only in 1908.

The incredible longevity of the site

Right: looking at Stone-Age rock carvings

makes it all the more fascinating: its span of creativity stretches from the Stone Age into Roman times. Ancient Rome's graffiti artists covered their predecessors' work with Latin inscriptions of the 'I woz here' variety.

Unless you need treatment for liver and digestive disorders, there is not much to detain you in **Boario Terme**, a shabby, formerly genteel spa resort 12km (7½ miles) north of Pisogne. Far more interesting is the mock Stone Age village of **Archeopark** in Place Gattaro (9am–5.30pm; tel: 0364-529552). This educational theme park near Boario Terme features lots of demonstrations of Stone and Iron Age activities as carried out by the Camuni people of 10,000 years ago. Children in particular enjoy sampling life in primitive rock shelters or relatively sophisticated stilt villages. After corn-grinding in the stilt village and tending to the smelly wild boar comes rowing in Stone Age-style flat-bottomed boats on the lake.

Hunting Skills

Less popular are the sections on the cultivation of flax, millet and lentils, and the functioning of Iron Age carts, axes and sickles. Workshops cover hunting and gathering skills, as well as archery, goat-herding, corn-grinding, copper-beating, stone-carving, spinning, flax-making and hide-tanning. School parties can book two-day courses, featuring Neolithic fishing (with bare hands) and bed-making from dry leaves and animal pelts.

A little way north, **Bienno**, which is dominated by a medieval castle, was known as a mining centre that survived from the Iron Age until World War II. In **Breno**, the modern valley's commercial capital, the Gothic **Chiesa Sant'Antonio** is adorned by Romanino's realistic frescoes. Midway between Breno and Capo di Ponte, **Cerveno** is home to the curious **Santuario**

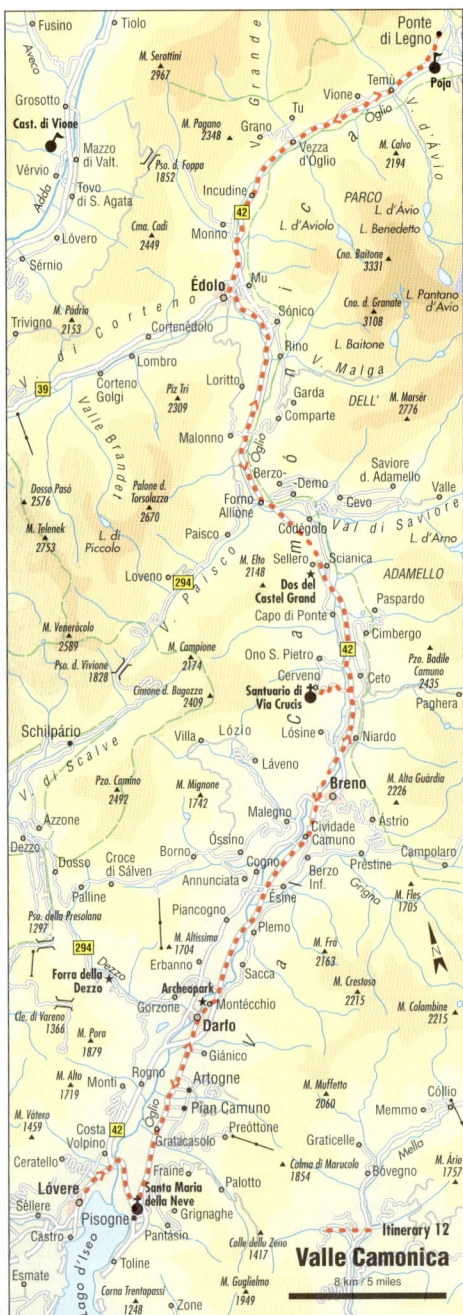

Valle Camonica

Itinerary 12

8 km / 5 miles

di Via Crucis. This much-loved church displays a distinctly kitsch *Stations of the Cross*, with 200 life-size 18th-century statues evoking the Passion.

Thousands of Carvings

In the heart of the upper valley, **Capo di Ponte** marks the main entrance to the **Parco Nazionale delle Incisioni Rupestri** (Tues–Sun 9am–1 hour before sunset; mornings only in winter; tel: 0364-42140). Given that at least 300,000 carvings are etched onto the glacier-seared sandstone of this, the national rock-engravings park, it is a good idea to focus on the impressive boulders near the entrance. Here you will see representative scenes spanning thousands of years, from Stone Age scratchings to Bronze Age narratives.

Animals, both symbolic and real, feature prominently. These range from Etruscan boxer dogs masked as cockerels to elk speared by hunters, and deer caught in lassos. Rock 1 is humorously entitled: 'When food is also a god', referring to the dual role of the deer as sacred symbol and venison meal. Nearby, the 'horsemen of the rocks' feature shows how horses were a common symbol of status. ('A horse then was like a Ferrari today,' as the guide says.) Although this open-air museum is linked by walkways, with numbered rocks and explanatory panels in English, you might opt for a guide, who can decipher a few of the tougher mysteries. Rock number 50 dates from 500 BC and features an Etruscan alphabet. The later Bronze Age scenes, depicting axes and daggers, give way to an impressive Iron Age portrayal of a horse-drawn wagon on rock 23. Even in the Iron Age, the engravers used stone flints.

For generations, Camuni hunters and farmers recorded everyday life and their relationship with the other world in scenes so complex that modern archaeologists remain baffled. Rock 32, probably selected for its soft, feminine contours, depicts a line of women but the meaning is obscure. Hunting, praying, dancing, copulating, invoking the gods, indulging in

Above: at the national rock-engravings park
Left: something moved down there

sacrificial rites: all human life is here, but the key is lost somewhere in these wild chestnut groves. Ploughing scenes are mixed with propitiatory rites, and images of warfare and weaponry, all interspersed with symbols of roses, solar discs and labyrinths. While opinion is divided as to whether or not the discs represent shields, the rose – probably the oldest representation of this flower anywhere in the world – is thought to symbolise sacrifice and the immortality of the Camuni people. As for the labyrinth, it might well represent the passage from this life to the next. This theory is supported by the frequent presence of the phoenix symbol, which was often depicted right beside it.

The wild scenery provided by the area's birch and pine woods appealed to the sensibility of the ancient Camuni tribes, whose attachment to natural beauty can be seen in their appreciation of the rocks. The almost palpable magic of the site is enhanced by the mysterious symbolism at the heart of the valley. Although clearly a sacred place, the site has not thrown up any graves so the supposition is that bodies were cremated and the ashes scattered. The mystique of the site has worked its way into the local consciousness – apart from the rose symbol, which has been adopted by the region, the valley is rife with legends of sacrificial virgins.

Once you have had your fill of prehistoric mystery, settle for a late lunch at Il Riccio *(see Eating Out, page 78)*, a friendly farm in the Campolungo countryside, some 5km (3 miles) north of Capo di Ponte.

Life in the Trenches

If you would like to make a day of this itinerary, continue up the valley, through Edolo to **Temu**, where you can visit the sombre **White War Museum** (5.30–7.30pm) on Via Adamello. The museum's somewhat fusty collection does justice to the Italians' struggle against Austro-Hungarian forces in the glaciers and snowy peaks of the Adamello Dolomites that formed a front line in World War I. In 1915–1918, bitter trench warfare dominated the snow-fields above the valley, and the museum displays uniforms and weapons from the front, mixed with poignant details of daily life in the trenches. Well-preserved weapons are still regularly unearthed on the glacier.

Before heading back to the lake, you might want to press on to the uplifting winter ski resort of **Ponte di Legno**, which is an equally pleasant alpine town in summer, not least due to the presence of any number of lively cafés.

Lake
Garda

13. SIRMIONE AND A LAKE GARDA CRUISE
(see map, p62)

A walking tour of medieval and Roman Sirmione, Lake Garda's loveliest resort, followed by a cruise as far north as Malcesine or Limone.

In Sirmione, park on the southern side of the peninsula (separated from the pedestrian area by a moat and busy during the Friday market). Check events, ferry times and dinner cruises at the tourist office (tel: 030-916114). For a cruise, take a ferry on the outward leg, and a fast hydrofoil for the return.

Set on a delightful peninsula at the southern end of Lake Garda, **Sirmione** enjoys a gorgeous location. The Romans were drawn to the invigorating waters around Lake Garda and, impressed by Sirmione's hot, sulphurous springs, developed the spa as a sybaritic retreat. The conquering Scaligeri counts from Verona were impressed by Sirmione's military potential and built a medieval fortress from which to govern the southern part of the lake.

Sirmione was for a long time subject to Venetian rule (until the end of the 18th century) and, in spite of recent over-commercialisation, it retains much of the architectural grace of that period.

The peninsula's lower, landward side is not particularly interesting but it is convenient for cafés, parking and tourist information. The 13th-century **Rocca Scaligera** (9am–6pm), a fort crowned by swallow-tailed battlements and encircled by water, guards the entrance to the historic centre: to enter Sirmione, you must cross the drawbridge over the moat.

Pageants and Concerts

In addition to the picturesque moat, the castle features well-preserved bastions and crenellations, and a fortified dock. In summer the Rocca stages pageants and concerts in the two castle courtyards. It also hosts temporary exhibitions to boost the appeal of the permanent collection of minor Roman and medieval archaeological finds. From the night watchmen's walk the urban views of the densely constructed medieval centre fade into olive groves and the lush, Mediterranean tail end of the peninsula. The area that encompasses the medieval nucleus, the Roman baths and the modern spa starts on the other side of the drawbridge and stretches to the end of the peninsula.

Via Vittorio Emanuele, the main street, leads northwards, with a right

Left: the undulating countryside of the Parco Nazionale delle Incisioni Rupestri
Right: the 13th-century Rocca Scaligera fortress in Sirmione

turn to the 15th-century church of **Santa Maria Maggiore**, which overlooks a tiny beach and lakeside walk. The frescoed church's portico incorporates a Roman capital. The surrounding web of tiny alleys abounds in cafés, over-priced art galleries, and chic, hand-crafted jewellery outlets. This engaging tourist trap is relieved by lake views and clumps of palms or parasol pines.

Via Emanuele continues northwards to the modern spa centre of **Terme** (tel: 800-802125; tel: 030-91681). Here the hot sulphur springs – bubbling radioac-tive waters are channelled up from the bottom of the lake – are utilised in the treatment of an assortment of respiratory complaints. Via Punta Staffalo runs from here to the western shore, with Via San Pietro, a turning to the north, lead-ing to **San Pietro**. This Romanesque church was constructed on top of the remains of a Roman temple, and was remodelled with recycled Roman bricks.

At the end of the promontory, medieval Sirmione can be explored to its Roman core. Just beyond San Pietro the revamped **antiquarium** (daily 9am–6 or 7pm) displays Roman finds, from fragments of frescoes and mosaics to coins and ceramics. This museum marks the entrance to the **Roman ruins**, including the original spa and villa complex. The Mediterranean mood is enhanced by the fragrant scent of rosemary, and the sight of sun-baked Roman bricks and olive groves. Paths wind past rocks and ruins, with views over the jagged shoreline.

Catullus Lived Here

The **Grotte di Catullo**, which crowns the rocky tip of the peninsula, is infused with heady romance, not just because the villa belonged to Catullus. Rome's greatest lyric poet languished here when rejected by Lesbia, his mistress in Rome. Once reached via a triumphal arch and barrel-vaulted arcades, the villa is now an atmospheric heap of stones. A geometric puzzle, it reveals a complex interplay of passages and porticoes, a sensitive blending of brick and rough-hewn stone.

To make the most of the cruise, have a light lunch at one of the first ports, such as Lazise. The classic cruise calls at Peschiera, Lazise, Bardolino, Garda, Gardone Riviera, Malcesine and Limone, but if you visit several resorts or have a leisurely lunch, head back to Sirmione from Garda. The best resorts, with medieval Scaligeri castles or elegant Venetian customs houses, are on the lower part of the lake, the eastern shore, in the Veneto.

The equivalent stretch of western shore, in Lombardy, is more mundane, although north of Salò, the elegant resorts are matched by a gloriously rugged landscape. Thanks to Lake Garda's exotic climate, the entire coast has its share of magnolias and oleanders, citrus fruits and conifers, olives and vine-yards, parasol pines, plane trees and palms.

Above: Grotte di Catullo, the home of ancient Rome's greatest lyric poet

Peschiera, east of Sirmione, is a historic port and military stronghold at the mouth of the River Mincio. The ferry then calls at the Scaligeri base of **Lazise**. As a former customs post and Venetian military outpost, Lazise reflects the region's colonial history. The Romanesque church of San Nicolo and the lively waterfront cafés and *pizzerie* make this a pleasant place for lunch, possibly in Bastia on Via Bastia. Just north is **Bardolino**, framed by rolling hills and vine-clad slopes. The distinctive reds of this wine-growing region can be tasted in any of the local bars. Romanesque churches, a ruined castle and a medieval quarter are particularly popular with German and Austrian visitors. Next comes **Garda**, a former fishing village sheltering in the lee of Mount Garda.

From Garda, the ferry crosses to **Gardone** *(see Itinerary 14, page 63)*, a prestigious resort on the Lombardy bank, and passes the loveliest stretch of coast, wild alpine terrain that stretches all the way to Riva del Garda *(see Itinerary 15, page 65)*. Despite the towering cliffs, the area is a hothouse for Mediterranean shrubs and citrus fruits as a result of the balmy microclimate. Citrus fruits, introduced by medieval monks, used to constitute the lake's cash crop. To protect the lemons from rare but catastrophic cold spells, the terraces were traditionally south-facing, covered with wooden supports in the colder months, and watched over by conscientious gardeners who would light fires if the temperature dropped suddenly. Citrus cultivation went into an irreversible decline in the 19th century, but these distinctive lemon terraces still line the lake from **Gargnano**, north of Gardone, to Limone.

A Family Resort

Further north, on the Veneto shore, the ferry stops at **Malcesine**, amidst the olive groves on the slopes of Monte Baldo. The loveliest resort on the east shore, with a café-lined waterfront, balconied Venetian-style houses and a maze of cobbled alleys, Malcesine is a worthwhile port of call. A quintessential family resort, it is popular with British visitors who appreciate the local charm, good hotels and water sports. There are good views of the main attraction, a crenellated castle that soars over red-tiled rooftops, from any lakeside bar.

The ferry returns to the Lombardy shore and the quaint resort of **Limone**. Less self-consciously cute than Sirmione, Limone is caught between lush Mediterranean vegetation and sheer rock-faces cowering under snow-clad

Above: exclusive Gardone
Right: Malcesine, a family resort

peaks. This old fishing port (beloved by D.H. Lawrence, Goethe and Ibsen) is a tad touristy – for all its bright southern light and pastel facades overhung by flower-bedecked balconies, the service is poor, the souvenirs tacky. Head back to **Sirmione** in a hydrofoil. Unless planning a dinner cruise with a stop at **Desenzano**, the resort with the liveliest nightlife, take a taxi home.

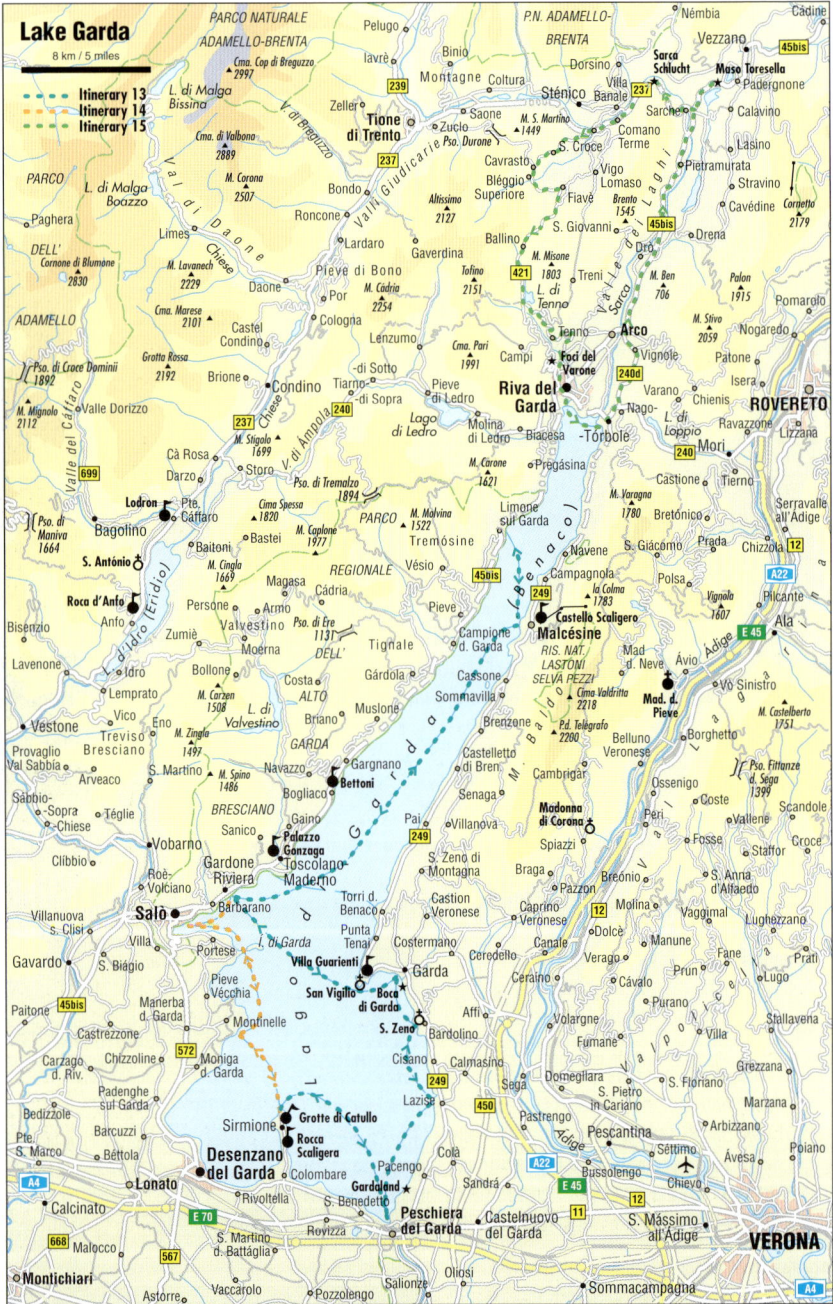

Lake Garda

8 km / 5 miles

- - - Itinerary 13
- - - Itinerary 14
- - - Itinerary 15

14. GARDONE RIVIERA   *(see map, p62)*

Sail or drive to Giardino Hruska botanical gardens and Il Vittoriale, the lakes' weirdest villa complex. Catch the ferry to the historic town of Salo.

From Sirmione, regular ferries travel to Gardone Riviera on Lake Garda's western shore. Book lunch or dinner in Villa Fiordaliso (tel: 0365-20158).

Just off the lakeside promenade, **Giardino Hruska** (9am–6pm) is a collection of lush botanical gardens created by Arturo Hruska, dentist to the czar, in 1912. The pleasant grounds feature an English garden full of tropical plants, a Japanese garden of pools and bamboo, and an alpine, Dolomite garden with landscaped waterfalls and ravines chiselled out of the rocks. A left turn after the park leads to a bizarre villa secluded by cypresses and oleanders.

The Dictator and the Poet

Il Vittoriale (Oct–Mar: 9am–5pm; Apr–Sept: 8.30am–8pm) is a testament to the megalomania of two men. Gabriele D'Annunzio (1863–1938), soldier, poet, fascist, aviator, aesthete and womaniser, occupies an odd place in Italian hearts, somewhere between reverence and bafflement. 'Destiny calls me towards Lake Garda', he declared, though it was actually Mussolini who presented him with the villa in 1925.

Disillusioned with the paltry gains won by Italy in the post-World War I peace – the Dalmatian town of Fiume (Rijeka) on the Adriatic had been promised to Italy but was presented to Yugoslavia instead – D'Annunzio and his private army occupied Fiume. Forced to withdraw in 1921, D'Annunzio retired to paint his gilded cage on Lake Garda while Mussolini pursued his own myth-making on the world stage. Not that the decadent aesthete and the brutal dictator were exactly soul mates. On one famous occasion D'Annunzio brazenly forced his fellow fascist to read the following inscription, which he had placed over a mirror: 'Remember that you are made of glass and I of steel'. Mussolini's reaction is not recorded but D'Annunzio lived to tell the tale.

Named in celebration of Italy's victory over Austria in 1918, and remodelled by D'Annunzio, the 18th-century Il Vittoriale is one of Italy's most flamboyant pre-war estates. The house, which can be seen only on a guided tour, features a cool reception room reserved for disliked guests, who included Mussolini; favourites were welcomed in a warmer chamber. D'Annunzio's delusions of grandeur led him to create a low entrance to his study so guests had to stoop, presumably to bow. The decadent decor raids sacred and profane motifs, with walls and

Above: Giardino Hruska botanical gardens date back to 1912
Right: a typical adornment at the flamboyant estate of Il Vittoriale

raids sacred and profane motifs, with walls and ceilings studded with crests, arcane symbols and secret mottoes. Islamic plates jostle for space with Austrian machine guns. D'Annunzio abhorred daylight so the windows were made of stained glass or painted over. When the penumbra became too much to bear, D'Annunzio would retreat to the coffin in the spare bedroom. His blue bathroom is filled with Moroccan trappings such as trunks of questionable ceramics. In the dining room, his embalmed pet tortoise, which died of indigestion, served as a reminder of the wages of gluttony.

Relics of the Fiume Fiasco

The splendour of the **grounds** contrasts with the ugliness of the creations that inhabit them: a magnolia grove houses a war memorial while the *Puglia* ship that featured in the Fiume fiasco is bizarrely beached among the cypresses. In a hangar are a biplane that flew over Vienna in the war, vehicles that took part in the Fiume debacle, and the Italian flag. The **mausoleum**, where Fiume casualties are buried, features D'Annunzio's kitsch, self-aggrandising tomb, and, displayed in an an eerie museum, his death mask. Still, pockets of the gardens are less oppressive: a lemon terrace has been transformed into a private garden and the fascistic amphitheatre has fine views from the top tier.

You could dine in Gardone's upscale Villa Fiordaliso *(see Eating Out, page 79)*, an Art Nouveau villa associated with Mussolini and his mistress. Or take

the ferry back to Sirmione via **Salò**, Mussolini's 1943–1945 seat of power. Salò, a town of fleeting moods rather than awesome sights, wears its history lightly. After an earthquake in 1901 the resort was rebuilt in airy Art Nouveau style and is still graced by elegant villa hotel-restaurants, such as Laurin. In summer there are often open-air concerts in Piazza del Duomo, where you will find a Venetian Gothic cathedral that survived the quake.

Above: Il Vittoriale's *Puglia* amid the cypresses
Left: interior of D'Annunzio's gilded cage

15. A TASTE OF TRENTINO *(see map, p62)*

An early-morning stroll around Riva del Garda's lakeside resort and castle, a ferry ride to Torbole and a dramatic drive via Arco to Toblino Castle for lunch. Return to Riva along back roads, passing a string of medieval hamlets. Dinner in Riva, followed (in summer) by a cruise.

Park by the castle, ideally in the car park in Giardini di Porta Orientale or the adjoining Congress Centre. Book lunch at Toblino Castle (tel: 0461-864036, closed Tues). In season take a cruise (tel: 0464-554444) from Riva.

Riva del Garda is Trentino's gateway to the largest inland lake in Italy, and the castles of Arco, Drena, Toblino, Stenico and Tenno are all a short drive away. As a medieval port for powerful prince-bishops, Riva became a pawn in the dynastic struggles between such city-states as Milan, Venice and Verona. In 1703 the port was sacked by the French during the War of the Spanish Succession, leaving Riva a shadow of its former self. Like the rest of Trentino, Riva was revived under Austrian rule (1815–1918), and it flourished as a fashionable resort, attracting such *mitteleuropean* literary heavyweights as Kafka, Nietzsche and Thomas Mann. Contemporary Riva del Garda is touristy but it retains a sleepy charm. Out of season the resort is popular with a largely middle-aged clientele; summer brings a younger crowd drawn to water sports as much as culture.

Moated Castle

The resort is centred on the lakeside **Rocca**, the moated medieval castle. This austere military stronghold, which once included arsenal, barracks and palace, has recently been restored but it evokes only a partial sense of its former glory. The Rocca was designed as a fortress but the Renaissance prince-bishops of Trento turned the interior into a gracious patrician residence. It was further domesticated in Austro-Hungarian times, when its fearsome appearance was compromised by the lowering of its corner towers. Beyond the drawbridge the revamped and child-friendly **Civic Museum** (9.30am–6pm; tel: 0464–573869), which has a minor art collection, is best visited at night. In the summer, evening events in the castle courtyard complement the collections. For instance, baroque music and dancing recall castle life under Venetian rule, a mood well-illustrated by the period portraits.

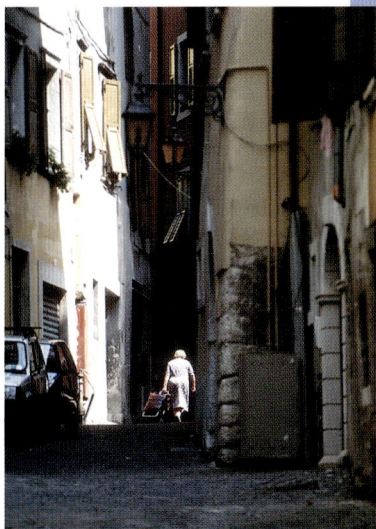

After the Rocca, you might hit a lakeside bar or, from the adjoining Piazza Garibaldi, take Via Mazzini into **Piazza Cavour**, the main inland square, which is often obscured by market stalls. You can buy local olives, cheese and wine at the Wednesday market; to sample Riva's best coffee, head for the traditional Caffè Maroni on the same square. The alleys around Piazza

Right: a back street in Riva, which was once fashionable with the European literati

Cavour form the heart of the shopping district and sell quality clothes and leather goods. For fine views take a short steamer ride to Torbole.

The **Piazza 3 Novembre** waterfront west of the Rocca is lined with 15th-century Venetian–Lombard palaces, including the town hall. With its 13th-century gateway, a view of the 16th-century bastion above, and Hotel Sole, once an Austro-Hungarian rulers' residence, the square is a microcosm of Riva's history. The hotel has lost much of its grandeur but its sunny terrace is good for drinks and people-watching. To see the less genteel face of Lake Garda, take the first ferry one stop south to Torbole from one of the jetties beside the hotel.

Surfers from Bondi to Bavaria

Torbole, once a fishing village, is a lively water-sports centre that benefits from unusual wind conditions. Here Bondi Beach surfers mingle with their bronzed Bavarian rivals. Just before midday, the *Ora* southern wind whips down the lake and fills the sails of acrobatic windsurfers until early afternoon. On the approach to Torbole you might see free climbers, seemingly suspended over the lake in the rocky playground of Corno di Bo just south of the resort.

The fish restaurants in Torbole are as good as any on the lake (*see Eating Out, page 80*). Or, for lunch in Toblino Castle, stroll back to Riva along the waterfront path. A new footbridge over the River Sarca estuary means that walkers can follow the waterfront the whole way. The scenic path hugs a lakeshore teeming with sunbathers, sailors and wet-suited surfers. Trentino's section of Lake Garda is ecologically pure: speedboats are banned and windsurfers must stay at least 200 metres/yds from the shore. The 30-minute walk comes out at Riva's Congress Centre and public gardens, where you can pick up the car.

Toblino Castle (10am–6pm) is a 15-minute drive into the alpine hinterland, following Viale Trento until it joins the main Arco–Trento road leading north. **Arco**, once a patrician spa favoured by Austro-Hungarian grand-dukes, is known for its gardens and pleasant climate. The former archducal gardens overlook the wizened stump of Arco Castle, which looms above the resort. Perched on a rocky crag, the castle is flanked by cypress and olive trees and has views over the river as far as Lake Garda. A steep path winds up to the castle but the interior, for all its fragments of Gothic frescoes depicting courtly scenes, does little justice to the striking setting.

Continuing north to Toblino, there are views of Drena Castle (10am–6pm) on the

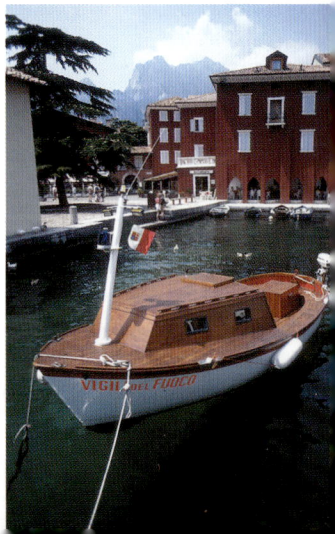

Above: café-bar facades in Torbole
Right: moored in Torbole

lake garda

right and of the glacial Lake Cavedine beyond. **Castel Toblino** (tel: 0461–864036), set on the tiny lake of the same name, and with an atmospheric restaurant, seems to materialise from nowhere. In his trailblazing *Italian Alps* (1875), the mountaineer Douglas Freshfield wrote of 'the little pool of Lago Toblino, rendered picturesque by its castle, an old fortified dwelling defended landwards by crenellated battlements'. A 17th-century prince-bishop kept a mistress, Claudia, in the castle. After tiring of her, he apparently fabricated a tale of incestuous love between Claudia and her brother – a crime punishable by drowning. Despite reported sightings of a ghostly boat and two wraith-like figures on moonlit nights, it's likely that Claudia outlived the bishop.

A Renaissance Wine Estate

The Castel Toblino restaurant's tasting menus feature lake fish and game. Diners can visit the atmospheric interior, complete with Renaissance courtyard and 17th-century stoves. You might enjoy an apple strudel and glass of wine in the bar (9am–10pm), the terrace of which overlooks the lake. Beyond the next bend is the fairy-tale Renaissance wine estate of **Maso Torresella**, founded by the prince-bishops of Trento, but now the seat of a more democratic consortium of growers. Dry white Nosiola wine and sweet Vino Santo dessert wine derive from the surrounding vineyards' grapes.

The leisurely country drive back to Riva takes minor roads through **Comano Terme**, a thriving spa resort, and the rambling medieval villages of Bleggio and Tenno. The start of this western route skirts the Adamello-Brenta park, named after the Brenta rock formations, a Dolomite group marked by towering limestone pinnacles and surfaces bathed in an orange sheen. Toblino's vineyards give way to neat barns and lush, alpine pastures, deep forests and sheer rock faces. Yet on the approach to Riva, the reappearance of olive groves represents the clearest shift back to the sultry Mediterranean microclimate of Lake Garda.

From Toblino, retrace your steps south and take the first right turn, towards Comano Terme. Now gleaming and modern, the spa has been valued since antiquity, with its curative waters gushing out of the rocks at 28ºC (82ºF). If you don't want to stop at the spa for treatment, continue south on roads marked to the medieval hamlet of Bleggio Superiore and, mired in the bogs, the archaeological site of **Fiave**. The Fiave nature reserve features the remains of a late Neolithic village. Further south lies **Lake Tenno**, an emerald-tinged lake formed by a landslide and ringed by a thick forest of pines and firs. The medieval hamlet of Tenno, clustered around a private castle, is several twisting kilometres north of **Riva del Garda**. Head for one of Riva's restaurants that overlook the illuminated Rocca *(see Eating Out, page 79)*.

Right: back to the beach

Milan

MILAN *(see map, p70)*

A day excursion to the historic city of Milan, visiting the cathedral, shopping for designer goods and relaxing in an old-fashioned pastry shop. Travel by train from one of the main resorts.

From Como, Stresa, Bergamo or Varese, the direct train journey takes about an hour. From the Stazione Centrale, Milan's main station, catch the metro to the Duomo (cathedral). Consider booking a gourmet lunch in Bice (02-76002572), the classic spot for Milanese risotto and celebrity-spotting, or even a night at the opera (02-72003744; www.teatroallascala.org).

Milan, the dynamic business and design capital of Italy, could not feel further removed from the languid atmosphere of the lakes. Yet the metropolis presents the perfect urban antidote to a quaint lakeside resort and a succession of blandly pleasant boat trips. The lure for most visitors is the chance to indulge in designer shopping in Italy's fashion citadel. The quality, range and concentration of upmarket consumer goods make Milan the obvious choice for shopaholics. Fashion, accessories, iconic designer objects for the home and luxury foodstuffs represent the best buys, all available at far lower prices than outside Italy. The greatest bargains are to be found in discount stores (often called 'stock-houses') where last season's stock can be bought cheaply by shoppers prepared to wade through rails of unsuitable goods in un-chic storerooms. Even die-hard anti-consumerists cannot fail to appreciate the sheer extravagance of the visual display in the glitzy designer district.

Further temptation comes in the form of culture on an international scale. If you wish to take in an opera at La Scala, prior booking is essential; otherwise, Milan Cathedral provides the main cultural treat. The historic city centre is easily manageable for visitors, and the distances between the majestic Duomo and the designer shopping district are walkable.

The Heart of Milan

Yet at its sartorial best, Milan is a cosmopolitan city that puts on a show with panache. **Piazza del Duomo** is the historical heart of Milan, with the magnificent cathedral providing the backdrop. The severe square is dominated by the **Duomo** (7am–6.30pm), Europe's largest Gothic cathedral. This unfinished masterpiece was begun in 1386, and seamlessly blends Gothic, baroque, neoclassical and neo-Gothic styles. French-style flying buttresses and soaring pinnacles contrast with the excessive width preferred by native builders. From the roof, accessible by lift, there is a memorable view of the sacred *Madonnina*, the cit protector, soaring over Gothic spires.

Just off the square, *La Rinascente*, the city's most

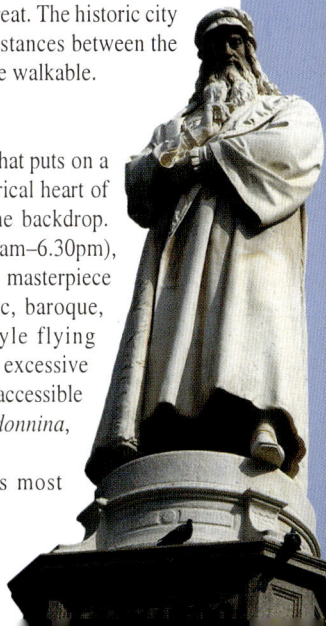

Left: Galleria Vittorio Emanuele
Right: Leonardo gazes down in the Piazza della Scala

upmarket department store, makes a possible first port of call for shopping, or for its panoramic rooftop restaurant overlooking the spires of the cathedral. If food shopping is on the agenda, then head for *Peck* (Via Spadari 7), a gastronomic temple just south of the cathedral.

Galleria Vittorio Emanuele, which connects the square with the opera house, is a splendidly arcaded shopping gallery known as Milan's front parlour. The myriad bookshops provide an intellectual antidote to the designer fashion dominating the city. Apart from elegant bars such as Salotto, Savini (closed Sat lunch and Sun, tel: 02-72003433) is the place for Lombard classics, but its formality can be oppressive. The Galleria leads to Piazza della Scala, home to **La Scala**, Italy's most celebrated opera house, which re-opens in November 2004 after a lengthy, and controversial, refurbishment. A first night here is an opportunity for Milanese matrons to take their furs out of storage and don the Bulgari jewels. Since Milan is Italy's design showcase, conversation soon turns from Verdi to the virtues of Gucci kitten heels versus Prada mules.

Fashion Rules

From the opera house, take Corso Matteotti to Piazza San Babila, the start of the **designer fashion district**, centred on **Via Montenapoleone**. Even if Milanese fashion victims can barely see through their Dolce &

Above: Milan's unfinished Duomo, Europe's largest Gothic cathedral

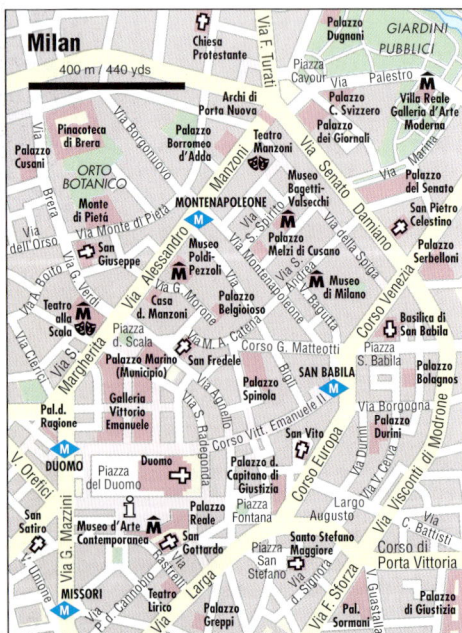

Gabbana sunglasses, they know where to find the best consumer durables. This so-called 'golden quadrangle' also encompasses **Via Sant'Andrea**, **Via del Gesù**, **Via Santo Spirito** and **Via della Spiga**. Here fashionistas swear by Fendi and Bottega Veneta for leather, La Perla for lingerie, Gucci for contemporary trousers, Armani for understated suits, Versace for glitzy evening wear, Missoni for knitwear, Moschino for shock value, Dolce & Gabbana for subversive sex appeal, and Buccellati for jewellery.

The main designers have several outlets on different streets in the district. A Prada flagship store at 8, **Via Montenapoleone** is awash with minimalist handbags. On the same street are Alberta Ferretti, Buccellati, Cerruti, Cartier, Dolce & Gabbana, Etro, Ferragamo, Frette, Gucci, Iceberg, La Perla, Pucci, Valentino, Versace, Vuitton and Ungaro. Ermenegildo Zegna, Jil Sander, Lacroix and Yves St Laurent are on **Via Verri** while Bottega Veneta, Bulgari, Dolce & Gabbana, Gigli, Krizia, Prada, Sergio Rossi and Tiffany await on **Via della Spiga**. **Via Sant'Andrea** has Armani, Chanel, Ferrè, Fendi, Hermès, Kenzo, Missoni, Moschino and Trussardi. **Via Manzoni**, which crosses Via Montenapoleone, is worth visiting for the Armani flagship store at number 31, which shows the shrewd designer's move into the lifestyle market.

Cova (Via Montenapoleone 8) is an old-fashioned café and pastry shop in the heart of the fashion district. This famous *patissier* has been producing *panettone* and *sachertorte* here since 1817. If a reviving coffee has convinced you not to spend your salary on a pink snakeskin belt, then consider the less chic shopping area of **Corso Buenos Aires** (metro Lima), a mecca for mid-market clothes, as well as for discounted designer garb from the previous season. Libero Outlet (Via Solferino 11) and Diffusione Firme Mode (Corso Buenos Aires 77) are typical designer discount stores.

The Brera Quarter

Later on, stroll along Via Brera north into the **Brera quarter** (metro Lanza), which is lined with galleries and bars. As night falls, **Via Fiori Chiari** fills with gypsy fortune-tellers and North African traders selling copies of designer handbags. The bohemian Jamaica bar (Via Brera 26) is the place to ponder on the likelihood of your peers recognising a fake while the Torri di Pisa (Via Fiori Chiari 21; tel: 02-874877) is a fashionable Tuscan restaurant. You can sample the vibrant nightlife in the canal district known as the *Navigli* (Porta Genova metro), or, if you are enamoured of the fashion district, then slink into the bar at the Four Seasons Hotel (Via Gesù 8). Set in a frescoed former monastery, this is a glamorous place for an early evening aperitif.

Right: bales of shimmering silk

shopping

Leisure Activities

SHOPPING

The lakes are awash with outlets selling seductive clothes, cult design icons, crafts, leatherware and gourmet gifts.

The recent advent of shopping malls *(centri commerciali)* has barely dented the popularity of small boutiques and department stores such as Rinascente and Coin. For classic clothes and designer goods, Milan *(see oage 69)*, Como, Verona, Varese and Brescia are the best bets. Como excepted, the lakeside resorts offer less choice and higher prices. Como, the 'city of silk', is also good for shoes and clothes. When it comes to crafts, Bellagio is good for hand-carved wood and glassware.

Discount Shopping

The region has a profusion of designer discount stores. In Milan, good stores include **Te Con Le Amiche** (Via Visconti di Modrone), **Diffusione Firme Mode** (Corso Buenos Aires 77) and **Il Salvagente** (16 Via Fratelli Bronzetti) where you can buy clothes by Comme des Garçons, Ferrè, Prada, Versace, Fendi, etc at a discount of 50 percent. **Vestistock** (Via Ramazzini 11; tel: 02-29514497) stocks Alberta Ferretti, D&G, Versace, Valentino, Ralph Lauren, Calvin Klein and DKNY for adults, as well as Armani Junior and Missoni Kids, all at up to 60 percent off normal retail prices. **Vesti Moda** (Via Manro Manhi 28; tel: 02-6707 1486) sells Armani, Versace, Ferrè, D&G, Alberta Ferretti, Etro and Valentino.

You can track down the main **Armani** outlet in suburban Milan (Via Provinciale per Bregnano 13, Vertemate con Minobrio; tel: 031-887373), reached by taking the Milan–Como train from Stazione Cadorna, getting out at Fine Mornasco, and then catching a cab to the Armani outlet 10 minutes' away.

The biggest and best discounts are arguably at **Serravalle McArthur Glen Designer Outlet** (80km/50 miles south of Milan or 50km/30 miles north of Genoa). It is just off the A7 motorway from Milan (exit Serravalle) and is open daily 10am–8pm (tel: 0143-686003; www.mcarthurglen.com). The 120 stores stock such brands as Prada, Bulgari, D&G, Versace, Diesel, Burberry, Ralph Lauren, Calvin Klein and Sergio Tacchini. There's also a bar, restaurant and children's play area.

The functional but well-designed **Franciacorta Outlet Village** is a good choice for those already visiting wine estates in the area *(see page 53)*. Sited in Rodengo Saiano, 7km (4½ miles) west of Brescia in the Lake Iseo direction, the shopping centre is also convenient for the A4 motorway from Milan (Ospitaletto exit). The village, which stocks such discounted brand names as Frette bedlinen, Versace, Guess and Benetton, is open from 10am–8pm, later on Fri and Sat (tel: 030-6810364; www.franciacortaoutlet.it).

Como's **Emporio della Seta** (Via Canturina 190; tel: 030-591420) is the place for discounted Como silk prints, fabrics, scarves designed by Ungaro, Lacroix and other top designers. **Ratti Sete** (Via Cernobbio 19;

Left: get your souvenirs here
Right: Milan, the home of *haute couture*

tel: 031-233262; open 9–noon, 2–6pm) sells discounted Como-produced silks for home furnishings, clothes, scarves and shawls.

Factory Outlets

Spacci (discount factory outlets) are also common around Milan, Franciacorta, Varese and Lake Maggiore. **Alessi** is the place for design icons – coffee pots, corkscrews and kitchenware. The Alessi showroom is at Corso Matteotti 9, Milan; the headquarters and factory shop are on Lake Orta (Via Alessi 6, Crusinallo; tel: 0323–6511).

Spacci Aziendali Riuniti (Via 42 Martiri 124, Fondotoce, Verbania, Lake Maggiore; tel: 0323–496545) sells name-brand pots and pans, cutlery and kitchenware.

Markets

Lake Maggiore hosts the region's biggest market at Luino (Wednesday); in Pallanza, there is an antiques and crafts market on the lakeshore on Friday evening.

On **Lake Como** there is a popular food and wine fair in Como (third Sunday of the month) and an antiques market at Varenna every third Sunday of the month.

There are lots of options in **Milan**, **Lake Iseo** and **Lake Garda**. Monday is market day in Peschiera and Torri del Benaco, Tuesday in Torbole, Desenzano and on the first and third Tuesday in the month in Limone. Riva's market is on the second and fourth Wednesday of the month, moving to Arco on the first and third Wednesday; Bardolino's market is on Thursday as is Lazise's, followed by markets in Sirmione and Garda on Friday, and in Salò on Saturday.

Food & wine

A pale, aromatic olive oil made around Lake Garda and Iseo is available from good supermarkets. You can buy local oil, cured meats and bottled peaches on **Monte Isola** at Alimentari Mazzucchelli. Superb cured meats are sold in **Domodossola** and in **Orta San Giulio** on Lake Orta. Grana Trentina cheese, sold in Riva del Garda, is a slightly inferior version of Parmesan but costs far less. For most lake produce try the **Riva Growers' Association** (10 Viale Luti; tel: 0464–552133), which sells olive oil, honey, jam, grappa and the finest regional wines.

Tourist offices in the wine-growing areas supply lists of estates (*cantine*) that offer tastings and sales. In Franciarcorta, near Brescia, recommended growers of *méthode champenoise* sparkling wine are Villa, Berlucchi and Longhi de Carli.

Above: a market comes to town
Left: delectable sweets and pastries

EATING OUT

Fish features prominently on most of the region's menus. In addition to lake trout, perch and prized carp, it includes lake sardines, eel, chub, tench and pike. The alpine influence can be seen in the array of cheeses, salami, polenta and mushroom dishes on offer in rural inns. The Austrian legacy around the north of Lake Garda has left the locals with a taste for veal, pork, beef, game, dumplings and gnocchi. Roasts, stews, game and white truffles are popular on the southern shore. Citrus fruits, olive oil, peaches and pears represent a Mediterranean input.

Fish dishes are generally washed down with white wines – especially Lugana, Pinot Grigio and Muller Thurgau – from Lake Garda, Trentino and Alto Adige. The sparkling wines (_méthode champenoise_) from the Franciacorta region near Lake Iseo are champagne in all but name: indeed their quality often surpasses that of the famous French brands. When it comes to red wine, the best regional offerings include Bardolino, Marzemino and Teroldego, as well as Franciacorta, which can be red, rosé or white. If you are in the market for one of the country's foremost wines, try a full-bodied Piedmontese Barolo, which is produced south of Lake Maggiore.

'Tourist menus' are best avoided, as are virtually all of the lakeside restaurants in the most popular resorts. In some of the best-known resorts, such as Stresa, there is a dearth of good restaurants outside of the top hotels. Standards are higher in touristy towns such as Bergamo and Verona.

Most upmarket restaurants require advance reservations, as do the rural ones, albeit for different reasons. Prices in the rural hinterland tend to be far lower than on the lakes, with anomalies in the fashionable wine-growing areas, such as Franciacorta, which is studded with upmarket inns.

Evening dinner cruises are recommended, though the mood, views and music might well be better than the cuisine. The same applies to most romantic island restaurants, with the exception of Iseo's Monte Isola, where it is difficult to eat badly in even the simplest lakeshore _trattoria_.

Price categories for a meal for two (with house wine) are as follows:

$ = Inexpensive (under 50 euros)
$$ = Moderate (50–100 euros)
$$$ = Expensive (over 100 euros).

Restaurant prices at the top end of the scale can vary tremendously. At prestigious restaurants, such as L'Albereta, you should expect to spend around 200 euros, before ordering wine.

Milan

Bice
Via Borgospesso 12
Tel: 02-7600 2572
A Milanese institution that serves traditional gourmet cuisine in a bustling atmosphere, and which is particularly popular with celebrities. Closed Mon and Tues lunch. $$$

Bistrot Duomo, Rinascente
Via San Raffaele 2
Tel: 02-877120
Far from being a bistro, this is a panoramic gourmet restaurant situated at the top of a department store. Enjoy Milanese and classic Italian dishes with views of the cathedral spires. Closed Sun and Aug. $$$

Boeucc
Piazza Belgioioso
Tel: 02-76020224/76022880
Occupying an old palace close to the Duomo, Boeucc offers classic gourmet cuisine in a gravely distinctive setting. Closed Sat and Sun lunch. $$$

Right: café life enjoyed alfresco

Premiata Pizzeria
Alzaia Naviglio Grande 2
Tel: 02-89400648
As its name immodestly suggests, Premiata Pizzeria is the foremost pizzeria in the city's cool canal quarter known as the Navigli. This is a particularly good option for travellers on a tight budget. Closed Tues. $

Lake Maggiore
Angera
Hotel Lido
Viale Liberta 11
Tel: 0331-930232
Angera is a recommended family-run fish restaurant situated in a lakeside hotel. Among the popular dishes are a platter of mixed grilled fish and seafood risotto. Closed Mon. $$

Rocca di Angera
Tel: 0331-931124/931300
Set inside the castle courtyard of the Rocca Borromea, Rocca di Angera serves highly creative cuisine. You can dine beside the 15th-century fireplace or on the castle terrace; both are atmospheric for lunch and dinner. Signature dishes include pumpkin risotto with truffles, and swordfish cooked with chicory. Also offers a good selection of classic regional wines. Closed Mon. $–$$

Laveno
Funivia
Top of the Laveno funivia (cable car)
Tel: 0332-660303/610303
An idiosyncratic chairlift deposits you at a restaurant that is blessed with superb views. Dishes include mushrooms, game and polenta, after which you might appreciate a walk around the summit. $

Menaggio
Vecchia Menaggio
Via al Lago 13
Tel: 0344-32082
A quietly reliable pizzeria, with good pasta too. Closed Tues, Wed. $

Stresa
Verbano
Isola dei Pescatori/Isola Superiore
Tel: 0323-32534
An atmospheric restaurant with a lakeside terrace that is especially fine in the evenings, when the island is quieter. Fish dishes predominate. If you are dining after 7pm, the restaurant will send its boat to pick you up from the mainland, usually Stresa, and then deliver you back again after dinner. $$

Verbania–Pallanza
Boccon di Vino
Via Troubetskoy 86
Tel: 0323-504039
A relaxed inn by the landing stage at Suno. Pasta, roast meats, cheeses and salami. Closed lunch, Sun & Aug. $

Lake Orta–Orta San Giulio
Ristorante Sacro Monte
Sacro Monte, Orta San Giulio
Tel: 0332-90220
An old-fashioned inn set in the sanctuary of Sacro Monte; the lofty lakeside views are complemented by rustic dishes. $

Ristorante San Giulio
Isola San Giulio
Tel: 0323-90234
Located on the main island on Lake Orta, this fish restaurant occupies an 18th-century villa with a terrace overlooking the lake. Prompt private ferry service back to the Orta San Giulio mainland. $

Above: Villa Crespi on Lake Orta is a luxurious Moorish fantasy

Taverna Antico Agnello
Via Olina 18, Orta San Giulio
Tel: 0322-90259
Gentrified rustic affair in the heart of town offers fresh local produce with a creative twist: home-made pasta, salami, game and risotto dishes. Closed Tues. $$

Villa Crespi
Via G Fava
Tel: 0322-911902
A luxurious Moorish fantasy situated at the resort's entrance, Villa Crespi does an imaginative blend of alpine and Mediterranean flavours, presented in sumptuous surroundings. $$$

Lake Varese
Vecchia Riva
Via Macchi 146, Schiranna, Varese
Tel: 0332-329300
Set on the banks of the unspoilt Lake Varese, Vecchia Riva has a sheltered garden and serves good fish, risotto and pasta dishes. The restaurant is attached to a small two-star hotel that offers functional modern rooms. $$

Lake Lugano
Du Lac
Viale Ungheria 19, Ponte Tresa
Tel: 0332-550308
Located in a small hotel on the Laveno road, just outside the port of Ponte Tresa, this lakeside restaurant serves good vegetable-based *antipasti* and risotto rather than the usual lake fish. $

Around Lake Como
Many of Lake Como's best eateries are in villa hotels *(see pages 94–95)*.

Argegno
Locanda Sant'Anna
Sant'Anna, on the Schignano road
Tel: 031-821738
A rural retreat in the hills above Argegno (15-minutes from the Como-Nord motorway exit). Local dishes – home-made *tortellini*, grilled vegetables, lamb, venison, wild boar, salami, alpine cheeses – are served in a garden setting with good views. The inn also has rooms *(see Accommodation, pages 94–5)*. $

Right: a big cheese, Isola Comacina

Bellagio
Silvio
Via Carcano 12
Tel: 031-950322
An unpretentious restaurant, Silvio benefits from a tranquil setting above the lake – not far from Villa Melzi – and a terrace. It also serves some of the freshest lake fish in Bellagio. A family-run hotel is attached to the premises. $–$$

La Pergola
Piazza del Porto, Pescallo
Tel: 031-950253
Situated in the tiny fishing port of Pescallo, a 10-minute walk through vineyards from Bellagio. This attractive inn specialises in inexpensive fish meals, which you can eat beneath the pergola. Accommodation (11 rooms) available. $

Como
Terrazzo Perlasca
Piazza de Gasperi 8
Tel: 031-300263
Trusty lakeside restaurant that is popular with the locals, Terrazzo Perlasca features a changing menu: classic fish and meat dishes as well as good pasta. Family-run. Closed Mon. $$–$$$

Sala Radetsky (Hotel Le Due Corti)
Piazza Vittoria
Tel: 031-328111
Charming restaurant in romantic setting in a former monastery by Porta Vittoria. Pasta, lake fish and local specialities. Closed Sun. $$

Spiaggia
Lido Villa Olmo, on the Cernobbio road
Tel: 031-570968
Fish restaurant convenient for Villa Olmo and the Lido, Como's beach. Closed Mon. $

Villa Geno
Viale Geno 12
Tel: 031-300012
www.villageno.it
Occupies a lovely neoclassical villa, with its own jetty. Serves classic Italian cuisine. $$

Isola Comacina
Locanda dell'Isola Comacina
Tel: 0344-57022
A huge set meal and a bizarre 'rite of fire' on a tiny island reached by boat from Sala Comacina. Open Mar–Oct. Closed Tues except in summer. Booking advisable. $$

Varenna
Vecchia Varenna
Corso Scoscesa 10
Tel: 0341-830793
Enjoy beautiful lake views and delicious fish. Closed Mon and in Jan. $$

Around Lake Iseo

Some of the best restaurants here are found in rural hotels (*see pages 95–96*) around Lake Iseo, Brescia and Franciacorta. Although close to Iseo, Franciacorta's excellent restaurants are accessible only by car/taxi.

Franciacorta
L'Albereta
Via Vittorio Emanuele II,
Erbusco, Franciacorta
Tel: 030-7760550
The region's top gourmet restaurant, whose menu relies on seasonal ingredients and the executive chef's whims. Italy's leading chef, Gualtiero Marchese, masterminds dishes that complement Franciacorta's sparkling wines. Booking essential. $$$

La Mongolfiera dei Sodi
Via Cavour 7, Erbusco
Tel: 030-7268303
Rambling former farmhouse that serves fine Lombard cuisine and Tuscan specialities served with Franciacorta and Tuscan wines. Closed Thurs and Aug. $$

Iseo/Clusane/Monte Isola
To eat well in this area, ignore the bland restaurants in the resort of Iseo in favour of any fish restaurant in neighbouring Clusane, in the hills, or on Monte Isola island.

Relais Mirabella
Via Mirabella 34, Clusane, near Iseo
Tel: 030-9898051
Recently established inn in the hills above Lake Iseo. Creative cuisine presented with a flourish, and a lovely romantic terrace. $$
   La Catilena, at the end of a path in the woods, is a traditional rustic-style restaurant owned by the same family. $

I Due Roccoli
Colline di Iseo, Polaveno
Tel: 030–9822977
Email: idueroccoli.com
This delightful hilltop eyrie above Iseo is the place for truffles, mushroom dishes and chocolate desserts. (Open to non-residents Thurs–Mon only.) $$

Bellavista
Siviano, Monte Isola
Tel: 030-9886106
This fine fish restaurant (and small hotel) specialises in grilled sardines, fish stew and mixed grills. Most of the island's fish restaurants are equally good. $/$$

Val Camonica
Il Riccio
Frazione Deria, Località Campolungo, Cedegolo
Tel: 0364-622113/0338-6083917
One of the best farm restaurants in Val Camonica, the friendly Il Riccio serves

Left: lakeside eateries abound

tasty dishes, including fresh pasta, chicken, wild boar, salami, cheeses, vegetarian dishes and forest fruits. It also sells its own produce. Booking essential. Open daily Mar–Dec. $

Bergamo
Taverna dei Colleoni e dell'Angelo
Piazza Vecchia 7
Tel: 035-232596
Situated in a gorgeous spot in the historic Città Alta – the upper town – the elegant Taverna dei Colleoni serves refined cuisine, especially fish dishes. Closed Mon and Aug. $$$

Brescia
Due Stelle
Via San Faustino 48
Tel: 030-42370
A long-established city-centre tavern (*osteria*), Due Stelle serves variants on traditional peasant cuisine, from an assortment of *antipasti* to chicken dumplings. Open Thurs–Sun all-day; Tues and Wed lunch only. $

La Vineria
Via X Giornate 4
Tel: 030-280477
An authentic wine bar and inn where you can sample excellent cold cuts and cheeses with recommended regional wines. Informal but informed tasting advice. Closed Mon. $$

Around Lake Garda
Brco
Alla Lega
Via Vergolano 8
Tel: 0464-516205
A well-established restaurant in a small resort in the Lake Garda hinterland, a short drive from Riva. Offers regional Trentino dishes from the lake and hills. Closed Wed and Feb. $$

Desenzano
Ristorante al Portico,
Via Anelli 44
Tel: 030-9141319
Classic fish restaurant, also serves good pasta. Closed Mon. $$

Gardone Riviera
Villa Fiordaliso
Via Zanardelli 132
Tel: 0365-20158
Small but stylish Art Nouveau villa hotel restaurant that was the home of Mussolini's mistress, Claretta Petacci. Stroll through the grounds after a gourmet lunch of risotto and lake fish. Book in advance. $$$

Malcesine
Vecchia Malcesine
Via Pisort 6
Tel: 045-7400469
Family-run gourmet restaurant tucked away on a scenic terrace in the old quarter. Highly flavoured dumplings and subtle pasta, plus classic lake dishes and Veneto wines. Book in advance. Evenings only; closed Wed. $$

Riva del Garda
Al Volt
Via Fiume 73
Tel: 0464-552570
Set in a historic palace and serving a mixture of dishes from Alsace and Trentino, Al Volt recalls the days of the Austro-Hungarian empire. Closed Mon, except in Aug. $$

Restel de Fer
Via Restel de Fer 10
Tel: 0464-553481
Set close to the prestigious Hotel du Lac, this gentrified rustic-style affair occupies a historic building, with summer dining in the cloister. Dishes inspired by old regional recipes. A handful of guest rooms. Evenings only, closed Tues. $$

Salo
Osteria dell'Orologio
Via Butteroni 26
Tel 0365-290158
Popular cosy inn serving lake fish as well as more bizarre dishes, such as skewered birds. Closed Wed. $$

Trattoria alla Campagnola
Via Brunati 11
Tel: 0365-22153
Traditional restaurant serving seasonal dishes using home-grown ingredients. Pasta and vegetables are prominent. $$

Sirmione
Osteria al Torcol
Via San Salvatore 30
Tel: 030-9904605
Cheerful inn and bar for ravioli, cold cuts and cheeses. Open 6pm–1am; closed Wed. $

Vecchia Lugana
Via Verona 71
Tel: 030-919012
One of Lake Garda's top restaurants, with lakeside terrace. Refined dishes feature fish and meat grills and seafood pasta. Wine-tasting in the cellar. Closed Mon, Tues. $$$

Spiazzo Rendena
Mezzosoldo
Via Nazionale 196
Tel: 0465-801067; fax: 801078
www.mezzosoldo.it
Family-run Michelin-starred village inn that serves only natural produce. The creative cuisine is based on meat, game, mushrooms and alpine cheeses. Book in advance. $$

Toblino
Castel Toblino
Toblino Castle
Tel: 0461-864036
A romantic castle by the lake, Castel Toblino offers a tasting menu that includes game as well as assorted fish from the lake. Prior booking is recommended on weekends. Closed Tues. $$

Torbole
Piccolo Mondo
Via Matteotti 7
Tel: 0464-50271
Trentino fare: game, dumplings, mushrooms, polenta. Closed Tues except in summer. $$

Ristorante La Terrazza
Via Pasubio 15, Torbole
Tel: 0461-506083
One of Riva's leading fish restaurants. Its specialities include chub ravioli, perch salad, lake sardines and grilled tench, washed down with crisp Pinot Grigio and Muller Thurgau. Very good value. Terrace. Closed Tues. $$

Dolomites Inns
Trentino and other alpine areas are known for their *rifugio* (literally 'refuge') inns, where you will find a friendly atmosphere, hearty meals based on regional peasant fare, and excellent value. Most are open in the summer and in the ski season, and are convenient for skiers, walkers and families.

Rifugio Al Cacciatore
Val d'Ambiez
Tel: 0465-734621
At a height of 1,800 metres (6,000ft) in the Brenta Dolomites, this alpine inn is reached by an invigorating three-hour walk along a clearly marked forest road from San Lorenzo in Banale – or book the jeep-taxi service. Big on mushrooms, polenta, salami and venison. Open 20 June–20 Sept only. $

Rifugio La Montanara
Pradel, near Andalo
Tel: 0461-585603
A simple inn with views over forests and unspoilt Lake Molveno. Access from Pradel via an old-fashioned cable car (or a steep 30-minute walk). Open May–Oct. $

Rifugio San Pietro
Monte Calino
Tel: 0464-500647
An inn near Canale with views over Lake Garda (access from Villa del Monte to Sella del Calino by car, then on foot for a 20-minute walk unsuitable for small children). Rustic fare features polenta, roast rabbit, risotto and spinach gnocchi. Open June–Sept. $

Above: simple pleasures
Right: a starting point for water-skiing

SPORT

Water Sports

Sailing is the region's main sport, followed by windsurfing and water-skiing. On Lake Maggiore there is sailing and windsurfing from Laveno and Luino, water-skiing on Lake Orta and rowing clubs on Lake Varese.

Como is a base for sailing, water-skiing, diving and canoeing. Water sports, including water-skiing and motorboat hire, are also on offer at Menaggio, Argegno and Bellagio.

Sailing is available from Pescallo, with windsurfing around Onno. Desenzano has the Fraglia Vela yacht club (030-9143343). Riva is the water-sports centre for the lake's north end. Riva and Torbole are the region's main windsurfing centres; Sirmione and Bogliaco are also good for windsurfing.

Motorboats are banned around Trentino, but are available elsewhere. You can hire rowing boats in Cernobbio, and there is canoeing on Lake Mergozzo, Lake Varese, and River Oglio. Rafting is popular in Val di Sole north of Lake Garda. *(For swimming, see Children's Activities, pages 82–3.)*

Golf

There are courses by lakes Varese and Maggiore. For clubs try Golf Tour Lago Maggiore (tel: 0321-620340; fax 0321-620814), an association of Maggiore clubs. There are clubs near Stresa, Verbania, Arona and Angera, but the best is Golf dei Laghi (tel: 0332-978101) near Lake Monate. For more information on golf, see www.stresa.net.)

Cycling

Lake Maggiore is opening cycle paths on the Piedmont side of the lake. There is a cycle path on one side of Lake Mergozzo, and routes around Menaggio and Bellagio. Trentino is good for mountain-biking: there are routes around Lake Garda, and lakeside paths between Riva and Torbole, where bikes can be hired (tel: 0464-506115).

Horse-riding

Horse-riding is popular on Lake Garda's mountain-bike routes. There are riding schools in Riva, Lonato (near Desenzano), Gargnano, Varese and around Lake Como.

Extreme Sports

Alpine guide associations or tourist offices can help arrange abseiling, paragliding, climbing and trekking. Cavalcalario, 3km (2 miles) north of Bellagio, is a good base for abseiling, paragliding and potholing. Monte Maddalena near Brescia is also good for hang-gliding and paragliding.

There is rock climbing around Lecco on Lake Como, and on Lake Garda, between Limone and Riva del Garda. Arco hosts the Rockmaster world championships in September. The Riva-Arco-Torbole axis is known for free-climbing – look out for climbers suspended above Lake Garda.

Winter Sports

Not far from the lakes there are a number of ski slopes, from Bormio – which is due to host the 2005 World Ski Championships (tel: 0342-902222; www.bormio2005.com) – to Turin, which hosts the 2006 Winter Olympics (tel: 011-6310511; www.torino 2006.org). Also Livigno, Pontedilegno, Tonale, Stelvio and Madonna di Campiglio. For top-quality summer skiing, head for Presena Glacier.

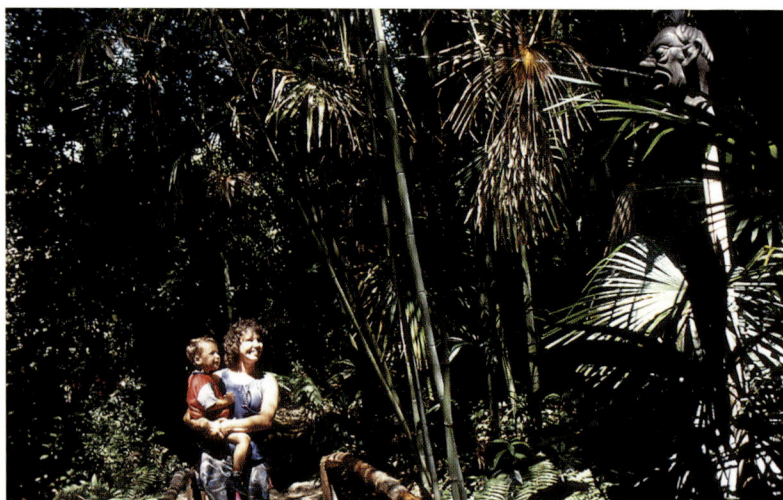

CHILDREN'S ACTIVITIES

The majority of the region's children's attractions, including theme parks and water parks, are situated at the southern end of Lake Garda. But there are plenty of other things that will appeal to children, from castles and cable cars to boat trips, bungee-jumping and swimming in the lakes.

Villas and gardens can be enjoyed by parents and children alike – the terraced gardens provide ample opportunity for rampaging, and Italians tend to be very patient when it comes to children. In northern Italy the journey is often more exciting than the destination, especially for children. The sheer range of possible boat trips – on old steamers, rowing boats, motorboats or ferries (with children under 12 travelling half-price) – should keep boredom at bay.

For a wonderful exploration of Swiss Locarno and the Lake Maggiore hinterland, take the highly scenic Domodossola Express from Stresa (summer only). A journey on this narrow-gauge railway can be combined with a Stresa–Locarno ferry trip.

Children also frequently enjoy cable-car rides: the best include Como to Brunate, Stresa to the botanical gardens, and Laveno to the peaks, with prodigious views over the Alps and Lake Maggiore. In the Dolomites you might take a cable car to alpine refuges, which are often both child-friendly (with playgrounds and swings) and adult-friendly (inns aplenty).

Theme parks & waterparks
Gardaland
Castelnuovo del Garda, outside Garda
Tel: 045-6449777; www.gardaland.it
Italy's answer to Disneyworld features a number of zones, including a magical kingdom, an Egyptian adventure, an African safari park and, the highlight for many youngsters, an exciting water park.

CanevaWorld
Lazise, Lake Garda
Tel: 045-6969900; www.canevaworld.com
Featuring the huge Aqua Paradise water park – designed to simulate a slice of the Caribbean, with fine white sands and palms – and a medieval section complete with fire-eaters and sword-fights.

Above: enjoying the botanical gardens
Left: one of many ice cream stops

Aquapark
*Riva Sport Centre, Viale dei Tigli 4/a,
Riva del Garda
Tel: 0464-552072*
The best place for water-slides and games in
Lake Garda. Also features a children's pool
and playground (daily 10am–6.30pm).

Acquasplash Franciacorta
*Via Dalla Chiesa 3, Corte Franca
Tel: 030-9826441; www.acquasplash.it*
Offers similar attractions to those of Aqua-
park, in the Franciacorta wine district near
Lake Iseo (daily 9.30am–7pm summer only).

Archeopark prehistoric theme park
*Place Gattaro, Boario Terme
Tel: 0364-529552*
Educational simulation of a Stone Age
village (daily 9am–5.30pm, closed Dec–Feb,
see Itinerary 12, page 56.)

Attractions

A visit to the **Alprose** Swiss chocolate
factory and museum (tel: 91-611 8856; daily)
in Caslano on the Lugano–Ponte Tresa main
road can be combined with a day trip to Lake
Lugano and the scenic resort of Morcote in
Switzerland.

The **Museo dei Trasporti** (Tues–Sun
10am–noon, 3–6pm) at Ranco on Lake
Maggiore is an engrossing presentation of
transport through the ages, with working
metros, trains, buses, a station and ski lifts.

Valeggio sul Mincio's **Parco Giardino
Sigurtà** (tel: 045-6371033; www.sigurta.it),
8km (5 miles) from Peschiera on Lake
Garda, features gorgeous gardens, a toy
railway, nature trails and cycle paths.

Parco La Quiete (tel: 030-9103171) in
Lonato, near Desenzano, is a pleasant park in
the hills with as swimming, fishing, mini-
golf and tennis courts, as well as playgrounds,
and barbecue and picnic areas by a small lake.

Rocca d'Angera on Lake Maggiore is
an imposing castle with impressive rooms
and views, and a doll museum (which might
be slightly creepy for young children).

Villa Pallavicino (daily 9am–6pm; tel:
0323-31533) in Pallanza on Lake Maggiore
is the setting for a villa, colourful gardens
and a children's play area, plus zoo with parrots, kangaroos, sea lions and zebras.

Beaches & pools
Lake Maggiore
The lake has clean waters, particularly on the
Piedmont shores, around Stresa, Verbania
and Baveno. Stresa has the Lido Blu and,
within easy walking distance, Spiaggia Lido
near Carciano – a sandy beach with a bar,
beach volleyball and boats to the islands.
Between Pallanza and Lake Mergozzo are
a cluster of good beaches, including the
(paying) sandy Lido di Suna (access via
Pallanza's Via Troubetskoy). Facilities
include showers, bars, and beach and water
sports, including volleyball, scuba-diving
and water-skiing. Just west are two free
pebble beaches with basic facilities.

Lakes Iseo and Como
There are good swimming centres for young
families near Iseo. Several have new pools,
a water park and grassy beaches leading
down to the lake. Try Lido Belvedere (Via
Colombera 15; tel: 030-980970), which has
canoeing and tennis, plus bars, cafés and
lakeside picnic spots. Elsewhere on Lake
Iseo, particularly on Monte Isola, the shore
suddenly descends into deep waters so swimming is safe for competent adults only. On
Lake Como, the water is cleanest some distance away from Como in areas such as that
around Bellagio.

Lake Garda
Between Riva del Garda and Torbole lies a
stretch of appealing, if often crowded, lakeside beaches, bordered by cafés and cycle
paths. This is the region's best windsurfing
spot (midday onwards). The windsurfers are
compelled to stay 200 metres (650ft) away
from the shore, and there are plenty of
lifeguards in attendance. The many
windsurfing learning centres
include one at Lido Blu *(see
Accommodation, page 97).*

Smaller Lakes
In Trentino, Lake Levico and Lake Lavarone
are popular with swimmers, and are
suitable for low-key water sports, such
as pedalos and boating. In Piemonte,
Lake Mergozzo is said to have the cleanest
water; Lake Monaco and Lake Lugano are
the cleanest in Varese province.

Right: shady parrot at the Villa Pallavicino zoo

CALENDAR OF EVENTS

If you are interested in attending any of northern Italy's internationally renowned cultural events, particularly the La Scala opera season in Milan, and the Arena di Verona summer opera festival in that city's Roman amphitheatre, advance booking is essential. For most of the folklore events booking is unnecessary, but check dates locally.

January–March

Carnival: (Feb or Mar).
Franciacorta Carnival: Allegorical floats parade in the village of Erbusco in the sparkling-wine country, and also in Bergamo, Varese, Bormio, Arco and, on the shores of Lake Iseo, in Pissogne and Clusane.
Colori e Sapori: The Franciacorta food and wine festival near Lake Iseo, including wine tastings (Mar or early Apr).
Milan fashion collections: Invitation-only.
Winter sports: There tends to be a world championships of some form every year in the region, including the world ski championships in Bormio in 2005 and the Winter Olympics 2006, held in Turin, Sestriere and other alpine resorts close to the lakes.

April/May

Camellia festival: Pallanza, on Lake Maggiore, stages events, walks and visits to appreciate the camellias in bloom on the Borromean islands and at key villas (Apr).
Tulip Week: Pallanza's Villa Taranto hosts an exhibition of gardens in bloom (Apr).
Ortafiori: Flower festival in Orta San Giulio on Lake Orta (end of Apr/May).
Mille Miglia: Italy's best-known vintage car rally, from Brescia to Rome and back, with concerts and events en route (May).
Cantine Aperte: Wine Day, celebrated by wine estates all over the region.
Villa San Remigio: Guided weekend visits to this romantic villa in Pallanza (until Aug).
BITEG: Food and wine fair in Riva del Garda (May).
VinArt: Celebration of wine and art, with shows, talks and tastings all over Trentino.

June

Arena di Verona: A major opera festival staged in Verona's Roman amphitheatre (until early Sept). Advance booking, tel: 045-8005151; book online at: www.arena.it.
Vinum Bonum: A Trentino food and music festival at wine estates around Lake Garda (June–Sept).
Rassegna gastronomica: A fortnight of regional food fairs around Lake Garda.
Estate Musicale del Garda: Summer concerts in atmospheric spots in the Gardone Riviera area.

Above: opera at the Arena di Verona
Left: celebrating Ferrari

Festa del Naviglio: Milan's summer festival of music, food and folklore on the canals.

Isola Comacina: Ancient water festival on Comacina, Lake Como (23, 24 June).

July

The Sounds of the Dolomites: 'Music in the mountains' free festival in Trentino. Musicians perform in open-air settings close to rural inns (until end of Aug).

Lago Maggiore Jazz: Lake Maggiore's most prestigious annual music event, with world-famous musicians performing in villas and castles in Stresa, Angera and at other resorts (usually in the third week of July).

Musica a Villa Giulia: Pallanza's free classical concerts by the lake (July and Aug).

Museum Nights: Trentino-based dance, music and historical re-enactments in museum venues (July and August).

August

Nightly **firework displays** over nine Trentino lakes.

Settimane Musicali di Stresa: Classical music festival in Stresa (third week of Aug).

Palio Remiero: Pallanza's night galleon race and firework display over the lake (usually in mid-Aug). Similar event in Laveno.

Palio delle Bisse Regatta: Lake Garda stages a gondolier-style rowing contest in memory of Venetian rule.

Notte di Fiabe: 'Fairytale Nights' for lake-side fun and fireworks at Riva del Garda, with similar firework displays at Limone, Salò, Sirmione and in Omegna (Lake Orta).

September

Festival of Sacred Music: Held in the monastery of San Pietro in Lamosa, near Iseo. International art exhibitions in Brescia's Museo Santa Giulia.

Corso Fiorito: Pallanza's biggest parade of floral floats, followed by fireworks.

Cento Miglia Regatta: prestigious sailing regatta starting in Bogliaco and racing to Riva, Sirmione, Desenzano and Salò.

Festival di Franciacorta: Sparkling-wine and food festival in Franciacorta.

Settimane Musicali di Stresa: Classical music in churches around Stresa.

Palio Baradello: Historical festival in Como with rowing and horse races.

Right: maestro Ricardo Muti conducting at La Scala Opera House

Italian Grand Prix: Formula 1 motor-racing in Monza, near Milan.

Traubenkur: A wine-tasting celebration north of Lake Garda..

October

Milan fashion collections: Invitation-only.

Pucia E Schisa: Barrel race in Erbusco, Franciacorta, plus food and wine fair.

Fiera dei Vini: Wine fair in Rovato, near Iseo (first weekend of the month).

Cantina in Cantina: 'From cellar to cellar' wine and food-tasting events in Franciacorta vineyards every weekend until December (tel: 030-7760870; www.strada delfranciacorta.it), and from March–June.

Food and wine harvest fairs, culminating in the Slow Food fair in Turin.

December

Christmas Fair in Como throughout month.

La Scala: 7 Dec marks the official opening of one of the world's most prestigious opera seasons, which runs until July in the newly restored opera house (tel: 02-72003744).

Presepi: Processions and Christmas cribs on display in churches throughout the region.

New Year's Eve son et lumière: Fun and fireworks at Riva del Garda.

ALISCAFO
A - PRENOTAZIONE

LUINO

CANNOBIO

BRISSAGO

ASCONA

LOCARNO

STRESA

Practical Information

GETTING THERE

By Air

The lakes are readily accessible from the revamped Malpensa airport (50km/31 miles from Milan), which is particularly convenient for destinations such as Lake Varese, Lake Lugano, Lake Maggiore, Lake Orta, and Lake Como. From the airport, take the fast Malpensa Express shuttle service (by train) to Milan.

Milan's Linate airport, conveniently located in the centre of the city, is well-served by quite a few low-cost and regular scheduled airlines, but most international flights now use Malpensa. Alternatives are Bergamo airport, which is ideal for Lake Iseo, and Brescia airport which, served by the low-cost Irish carrier Ryanair, is good for the western shore of Lake Garda.

Verona airport, situated at the eastern end of the lakes, is handy for the Veneto shore of Lake Garda, and for excursions to Brescia, Mantua and Venice. Lugano airport in Switzerland is another useful gateway to the Italian lakes, as are Venice and Treviso.

By Rail

Milan and Como are well-served by trains from Switzerland, Germany and France. If arriving from elsewhere in Italy, there are reliable connections from Turin, Bologna, Florence and Rome, and from within the lakes region *(see Getting Around, page 88)*.

By Road

Reasonable motorways *(autostrade)* and main roads link the lakes from Turin, Milan, Como, Varese, Bergamo and Brescia. But bear in mind that the city centres and ring roads tend to be both confusing and congested. New motorways on Lake Maggiore and Lake Iseo are improving the situation. Heavy traffic is now diverted to highways and tunnels above the lake, thus freeing the lakeside road for more leisurely journeys. From Milan to Lake Como, take the A9 Milano-Laghi motorway.

Left: departure times
Right: an opportunity to cool down

TRAVEL ESSENTIALS

When to visit

Spring, summer and autumn are all lovely times to visit the lakes, not least due to the fine weather conditions. If you want to see the blooming of camellias, azaleas and tulips in the lakeside gardens, April and May are usually constitute the best months.

During the liveliest period (June–end of August), overcrowding in the main resorts is offset by the pleasure of concerts, regattas and firework displays. September is a lovely month on the lakes, and the time of harvest festivals. October and November can be damp, but usually include quite a few unexpectedly fine days.

Weather

The lakes are where the Alps meet the Mediterranean, and as a result the climate can be unpredictable. Local variations depend on the size of the lake, and its degree of shelter. Shielded by mountains, the major lakes enjoy a Mediterranean microclimate. Lake Garda tends to be slightly colder than Lake Maggiore and Lake Como in winter, but slightly warmer in the summer. From July to October rainfall rates in the lakes are negligible compared with those of Milan, and the region suffers none of the city's stifling summers and foggy winters.

Average temperatures: Jan: Milan 2°C (36°F); lakes 5.5°C (42°F); Apr: Milan 13°C (55°F); lakes 13°C (55°F); July: Milan 25°C (77°F); lakes 24°C (75°F); Oct: Milan 14°C (57°F); lakes 14°C (57°F)

Clothing

Smart-casual clothes are the order of the day, but you will need something more formal at La Scala. Summer months excepted, it is worth dressing in layers. Take a jacket for summer evenings, when it can be cold.

Visas and Passports

EU citizens with a valid passport do not need a visa. Citizens of the US, Canada, Australia and New Zealand can stay for up to three months without a visa, and can extend their stay with a visa from the Italian embassy or consulate (easier done in advance).

Electricity

Italy has the same voltage as other European countries (220 volts) but you might need socket adapters for the two-pin sockets.

MONEY MATTERS

The Euro

The euro's replacement of the Italian lira led to a noticeable price hike. Travellers cheques are in euros, sterling or dollars. The euro is divided into 100 cents, with coins worth 1, 2, 5, 10, 20 and 50 cents. There are 1-euro and 2-euro coins. The seven euro bank notes come in denominations of 500, 200, 100, 50, 20, 10 and 5 euros. These notes are uniform throughout the EU's EMU (Economic & Monetary Union) nations.

Banks and Credit Cards

If you want to exchange currency, avoid 'Change' or '*Cambio*' places because they generally charge far more commission than the banks. In Italy the use of credit cards is less widespread than in the US and much of the rest of Europe – and sometimes there is an additional charge for credit card payments – so always check whether cards are accepted and whether payment by card will cost you more. Most Italian ATMs have instructions in the main European languages.

Sales Tax (IVA): Tax-free shopping is available for non-EU residents, with refunds given on production of a form provided by the retailer. IVA is 20 percent.

Tipping: Tipping is not obligatory but it is appreciated, even if a service charge is included. After good service in a restaurant, it is customary to round up the bill.

GETTING AROUND

Trains

Eurostar and InterCity trains are the fastest options. Book seats in advance (and pay a supplement). Tickets need to be validated before departure. To avoid queuing, buy your tickets from a travel agency. Milan, with its numerous stations, is the main rail hub for the lakes. Journey times are as follows:
• Milan–Brescia: 90 minutes
• Milan–Bergamo: 50 minutes
• Milan–Como: up to an hour
  The train trip to Lake Maggiore is pleasant: the Milan–Stresa train takes 65 minutes from Milan's Porta Garibaldi Station. The Domodossola Express follows a scenic route from Stresa and is a good way to explore Lake Maggiore's hinterland.
  On Lake Como, the most useful train service is via Como, which has efficient rail connections to Milan and Varese. Lake Varese and Lake Garda (using the station at Brescia) are well-served by trains. Journey times are as follows:
• Como–Varese: 70 minutes
• Milan–Varese: 70 minutes
• Brescia–Verona: 30–60 minutes.
  Desenzano, on the Milan–Venice line, is the main terminal on Lake Garda. The regular Iseo–Brescia mini-train takes 45 minutes.

Buses and Coaches

It can be quicker to reach some destinations by coach. For example the Brescia–Gardone journey takes only an hour by coach, which is also the best way to travel the Stresa–Lake Orta route. Also on Lake Garda, Sirmione, Salo, Gardone and Limone are inaccessible by train from Desenzano. For transport options around Lombardy, check out the www.regione.lombardia.it/trasporti website.

practical information

Ferries/Steamers

Ferries *(battelli)* offer the most enjoyable and leisurely way of exploring the lakes. Old-fashioned 'steamers' are more romantic but there is only one of these vessels on Lake Maggiore and another on Lake Como, and both sail in summer only. On the main lakes, ferries and catamarans cover similar routes.

Catamarans

Catamarans *(aliscafi* or *catamarani)* are fast but passengers are confined to the inside so they don't offer the views and atmosphere of a ferry. Be warned that catamaran and ferry stops can be at different jetties. Catamaran tickets allow you to use the ferries but not vice versa. On popular summer routes on the main lakes, the catamarans fill up quickly.

Boat Tickets and Timetables

There is a bewildering variety of tickets and deals for the major lakes so check the options before your first trip. These include: an all-day ticket for parts or all of the lake; a single or return ferry or catamaran ticket; and a ferry ticket that includes entry to (or a price reduction at) major sites. Some boats have a bar, others have restaurants, which often (especially on Lake Garda cruises) feature fixed-price, three-course meals.

All-day tickets generally cover ferry travel only, so in most cases you have to pay a supplement for the catamaran. This ticket is the most convenient option but, in terms of cost, it is worthwhile only for several journeys on the same day. Generally children under the age of 12 travel half price. Over-60s from EU nations are entitled to a 20 per-cent reduction on weekdays (with proof of identity). Return boat tickets (excluding catamarans) tend to be valid for two days.

Timetables tend to change at least twice a year, but the routes remain quite constant. For information and timetables, contact a local tourist office or ask at the ticket offices on the jetties *(imbarcaderi)*. Given the importance attached to meal times in Italy, you might find that services between 12.30 and 2.30pm are limited.

Cars can be transported on car boats *(motonavi)*, for instance on Lake Garda, between Desenzano, Limone and Riva.

The main boat companies are as follows:

Right: all aboard the *Stambecco*

Lake Maggiore
Navigazione Lago Maggiore
Via Baraca, Arona
Tel: 0322-232200; freephone: 800-551801; in Stresa, tel: 0232-30393; website (all lakes): www.navigazionelaghi.it
There are regular ferry services to Baveno, Verbania, Pallanza, Laveno and the islands. Frequent services connect Stresa, Pallanza and Laveno with the three Borromean islands. Catamarans connect the Italian resorts with Locarno in Switzerland, car ferries connect Intra and Laveno. There are also steamship cruises and evening cruises. For private boat hire, from water taxis to houseboats, call tel: 0323-933982 and 0322-7547.

Lake Como
Navigazione Lago di Como
Piazza Cavour, Como
Tel: 031-579211 or freephone 800-551801
In the lower and central part of the lake, there are frequent services from Como to Cernobbio, Bellagio, Tremezzo, Menaggio and Varenna, as well as services to the upper end of the lake.

Lake Iseo
The state railway runs to Iseo, from where there are ferries to Monte Isola and to the main towns, such as Lovere and Sarnico.

Lake Garda
Navigazione Lago di Garda
Piazza Matteotti, Desenzano
Tel: 030-914511 or freephone: 800-551801; in Riva del Garda, tel: 0464-552625
The main ports of call are: Desenzano, Sirmione, Peschiera, Lazise, Bardolino,

Garda, Moniga, Manerba, Salo, Gardone, Maderno, Bogliaco, Gargnano, Castelletto, Campione, Malcesine, Limone, Torbole and Riva del Garda. There are also request stops.

Motorboats

Motorboats are banned on the Trentino part of Lake Garda. Motorboat taxis aren't cheap – try to negotiate a day or evening rate.

Evening Boat Services

There is little public transport on the lakes after 8pm so if you want to dine in another resort, organise transport for the return trip. Bear in mind that private motorboats are generally far more expensive than taxis. Some island restaurants might transport diners back to the mainland. An inexpensive ferry makes the crossing to some islands: on Lake Orta the crossing is between Orta San Giulio and the island; on Lake Como the crossing links Sala Comacina with Isola Comacina. If you are worried about being stranded, discuss the options with your restaurant, or consider a dinner cruise.

Evening Cruises

Generally available in summer only, evening cruises range from dinner boats to discos. In most cases there is at least a bar and some form of entertainment, usually live music.
Lake Como: In addition to a 1926 summer steamer, dinner and dance cruises link Como, Bellagio, Menaggio, Verenna and Lecco.
Lake Maggiore: Navigazione Lago Maggiore *(see page 89)* runs a wide range of cruises. The steamship connects Stresa, Arona and Angera; the 'spaghetti cruise' serves pasta dishes. Several disco and dinner cruises link Luino, Laveno and Intra or Pallanza, Baveno, Stresa, Angera and Arona.
Lake Garda: Motonave Italia's summer cruises include evening cruises around Lake Garda. A typical trip picks up passengers at Riva del Garda, Limone and Malcesine between 8 and 9.30pm, and returns to these ports at about midnight. Another night service includes a disco boat that leaves and returns to Desenzano. The best gourmet dining cruises are on the 'Battello del Gusto' from Riva. For details call the tourist office in Riva (tel: 0464–554444) or Dezenzano (tel: 030–9141510).

By Car

It's quite feasible to rely on water transport while staying in the lakes region. If you are based on one lake, a car is unlikely to be necessary, even for visits to neighbouring towns or another lake. The convenience of a car is for touring: public transport between lakes can be tricky, and may involve going via Milan, which breaks the peaceful spell of the lakes. A car might be an asset if you are based on Lake Garda because north-south ferry crossings take so long. You will also need a car to follow the Franciacorta wine route, to visit the prehistoric sites north of Lake Iseo, or to dine in restaurants not in your resort. If travelling by car, check that the hotel has, or can recommend, a car park.

HOURS & HOLIDAYS

Business Hours

Banks are generally open weekdays 8.30am–1.30pm and 3–4.30pm; shops 9.30am–7.30pm. Many shops, especially larger tourist outlets, are increasingly staying open in the lunch hour, but many are closed 1–3.30pm. Resorts such as Bellagio and Stresa notwithstanding, most shops are closed on Sunday, and on Monday morning.

Public Holidays

1 January
6 January
Easter Monday
25 April (Liberation Day)
1 May
15 August (*Ferragosto*, the official start of the holiday season)
1 November
8 December
25–26 December

NIGHTLIFE

For most visitors to the lakes, entertainment is low-key: dinner in a romantic spot, a stroll along the waterfront, perhaps an evening cruise or a concert. If this sounds too sedate, you could head for the clubs and bars in Milan's Navigli canal quarter. Alternatively, consider a night in Desenzano on Lake

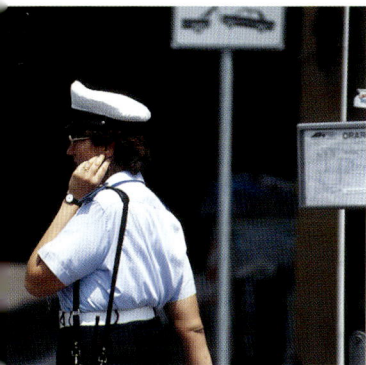

Garda, which is the focus for summer nightlife for the residents of Brescia and Milan. If you want to spend money in style, there is a casino in Campione d'Italia, an Italian enclave in the middle of Lake Lugano (passport and evening dress expected).

ACCOMMODATION

The official grading system is idiosyncratic – 3-star hotels can be more expensive than 4-star, depending on facilities. There can also be quite a range of standards within a single hotel, especially in the popular resorts. Ask in advance about features you might want, such as quietness, spaciousness and lakeside views (for which there may be an extra charge). If possible check out the room before you commit yourself. For a double bed, specify a '*camera matrimoniale*'. In high season, some hotels, especially in the most popular resorts, may be half-board only. A number of hotels in the lakes are closed in the low season (Nov–Mar).

Accommodation options are generally good in the lakes, certainly in terms of choice, range and setting. (Brescia, however, is full of soulless business hotels.) For accommodation in Milan, which generally represents poor value for money, call Milan Hotels Central Booking; tel: 02-805 4242; www.hotelbooking.com.

In Milan, if your stay coincides with one of the many trade fairs, the hotels will be full. Late July and the whole of August tend to be quiet periods, and even at weekends there are often vacancies. Many hotels may be slow to respond to email; reservations still

tend to be made largely by phone or fax.

The following price categories represent the average rate for a standard double room:
$$$ = from 200 euros (over 400 euros for deluxe suites)
$$ = 100–200 euros
$ = under 100 euros

Milan

Le Meridien Gallia
Piazza Duca d'Aosta 9
Tel: 02-67851; Fax: 02-66713239
www.lemeridien.com
This historic yet relaxed 5-star palace hotel near the Central Station has recently been revamped by Meridien; friendly, competent staff; huge marble bathrooms and elegant bedrooms; fine restaurant; famous bar; health and spa centre. $$$

Century Tower Hotel
Via F Filzi 25/b
Tel: 02-67504; Fax: 02-66980602
This welcoming, 4-star hotel occupies a tower close to the Central Station. All rooms are spacious suites with tea/coffee-making facilities; good restaurant. $$$

Spadari al Duomo
Via Spadari 11
Tel: 02-72002371; Fax: 02-861184
www.spadarihotel.com
A stylish 4-star hotel near the cathedral , Spadari al Duomo features spacious rooms, romantic works of art and flowers in public rooms, and a cheerful bar. $$$

Antica Locanda Leonardo
Corso Magenta 78
Tel: 02-4633717; Fax 02-48019012
www.leoloc.com
Stylishly restored, individualistic inn with inner courtyard and garden. Popular with the fashion crowd. $$

Antica Locanda dei Mercanti
Via San Tomaso 6
Tel: 02-8054080; Fax: 02-8054090
www.locanda.it
Charming boutique hotel with the cosiness of an old-fashioned inn, situated close to the Castello Sforzesco. Rooftop views from some rooms. $$

Above: a car might be an asset if you are based on Lake Garda, but watch out for traffic wardens

Palazzo delle Stelline
Corso Magenta 61
Tel: 02-4818431; Fax: 02-48519097
www.hotelpalazzostelline.it
Situated in a historic palace, the 3-star Palazzo delle Stelline is elegant, spacious and well-furnished. $$

Around Lake Maggiore
Angera
Lido
Viale Liberta 11
Tel: 0331-930232; Fax: 0331-930656
Engaging villa hotel with private beach. Rooms in this 2-star hotel are peaceful and functional. Good fish restaurant (try the ravioli filled with four types of lake fish). $

Baveno
Hotel Beau Rivage
Tel: 0323-924534; Fax: 0323-925253
Waterfront hotel in the heart of a quiet resort. An old-fashioned, family-run 3-star affair, it has freshly renovated rooms and top floor suites, a restaurant terrace, and gardens. $$

Laveno
Il Porticciolo
Via Fortino 40
Tel: 0332-667257; Fax: 0332-666753
Scenic 3-star waterfront inn with comfortable rooms and a fine fish restaurant that serves lake fish, trout ravioli, scampi risotto. Closed Tues. Hotel $; restaurant $$

Luino
Camin Hotel Colmegna
Via Palazzi 1
Tel: 0332-510855; Fax: 0332-501687
www.camin-hotels.com
An 18th-century hunting lodge set in a delightful spot, this family-run hotel retains its rustic charm. Restrained interior decor; invitingly individualistic rooms; restaurant terrace (grilled fish); waterfall; garden trails; seasonal rates. $$–$$$

Camin Hotel Luino
Viale Dante 35
Tel: 0332-530118; Fax: 0332-537226
www.caminhotelluino.com
Family-run, 4-star Art Nouveau villa; stylish bedrooms; good fish restaurant. $$

Ranco
Il Sole di Ranco
Piazza Venezia 5
Tel: 0331-976507; Fax: 0331-976620
www.ilsolediranco.com
Set 4km (2½ miles) north of Angera, this peaceful, family-run 4-star villa hotel is decorated in a graceful, restful style; terrace; lush grounds and lakeside views. Gourmet restaurant (closed Mon lunch and Tues). $$$

Stresa
Grand Hotel des Iles Borromees
Lungolago Umberto I, 67
Tel: 0323-938938; Fax: 0323-932405
www.borromees.it
Stresa's *grande dame*, this historic waterfront hotel faces the Borromean Islands. The *belle époque* decor is stylishly patrician if ponderous; palm-shaded gardens; beach; two pools, spa and fitness centre; tennis. $$$

Regina Palace
Lungolago Umberto I, 33
Tel: 0323-936936; Fax: 0323-936666
www.regina-palace.it
A grand Victorian affair with standards just below those of the Grand Hotel. It has similar facilities and an air of faded grandeur; noted Charleston restaurant; tennis; pool. $$–$$$

Verbano:
Isola dei Pescatori
Tel: 0323-30408; Fax: 0323-33128
Situated on a tiny island off Stresa, on the Borromean islands, this tiny romantic hotel is at its best when the hordes have gone; restaurant terrace; regular day ferries. $$

Villa Aminta
Via Sempione Nord 123
Tel: 0323-933818; Fax: 0323-933955
www.villa-aminta.it
Family-run luxury hotel in old-fashioned, flouncy style. Elegant restaurant. $$$

Varese
Palace Grand Hotel Varese
Via Manara 11
Tel: 0332-327100; Fax: 0332-312870
www.palacevarese.it
Varese's top hotel, an Art Nouveau palace in the hills. The public rooms have a ponderous

charm, suites are spacious. Guests include Bob Dylan and Catherine Deneuve. $$–$$$

Villa Castiglioni
Induno Olona
Tel: 0332-200201; Fax: 0332-201269
www.hotelvillacastiglioni.it
Set in parkland about 8km (5 miles) from Varese, the villa once sheltered the fugitive Garibaldi. Today it features canopied beds, original frescoes and ceramic floors. Has a refined restaurant with terrace. $$$

Bologna
Via B Broggi 7
Tel: 0332-232100
A welcoming 3-star, family-run hotel in the heart of the old town with spacious, simply restored rooms. Attached is a fine trattoria with specialities from Emilia Romagna, including pasta, salami and antipasti. Restaurant closed Saturday and August. $

Lake Lugano
Stella d'Italia
Tel: 0344-68139; Fax: 0344-68729
www.stelladitalia.com
A traditional waterside 3-star hotel located in the pretty village of San Mamete, with a restaurant, terrace and a jetty with private boat. Walks and horse-riding are available nearby. Apr–Oct. $–$$

Villa Cesarina
Ganna
Tel: 0332-719721; Fax: 0322-719007
www.villacesarina.it
An Art Nouveau villa overlooking tiny Lake Ghirla, in the hamlet of Ganna 6km (4 miles) from the Swiss border. A good point from which to explore the Italian side of Lake Lugano or Varese. Period style decor; convenient for golf and horse riding. $

Lake Mergozzo
La Quartina
Via Pallanza 20, Lake Mergozzo
Tel: 0323-80118; Fax: 0323-80743
www.laquartina.com
This family-run hotel is a lot quieter than many of the hotels on neighbouring Lake Maggiore. Its renowned restaurant overlooks the lake and grassy beach. The owner-chef

uses best quality local produce, such as lake fish and cheeses and salami from the Domodossola hinterland. $$

Lake Orta – Orta San Giulio
Villa Crespi
Via G Fava 8
Tel: 0322-911902; Fax: 0322-911919
www.lagodortahotels.com
A grandiose Moorish folly with atmospheric public rooms; personalised bedrooms with canopied beds and whirlpool baths; lovely restaurant with sophisticated cuisine. $$$

La Contrada dei Monti
Via dei Monti 10
Tel: 0322-905114; Fax: 0322-905863
www.orta.net/lacontradadeimonti
Set in a restored 18th-century palace with exposed beams, antiques, wrought-iron beds and original fireplaces. It lacks lakeside views and restaurant, but is good value. $–$$

Leon d'Oro
Piazza Motta
Tel: 0332-911991; Fax: 0332-90303
A family-run hotel overlooking the island of San Giulio. The bedrooms are a bit small and disappointing but most feature good views; noted restaurant with terrace. $$

Hotel Orta
Piazza Motta
Tel: 0322-90253; Fax: 0322-905646
Email: info@hotelorta.it
A historic hotel on the waterfront square. The maze of small, basic rooms is redeemed by charming terraces; reasonable restaurant; try to choose between several bedrooms. $–$$

Right: the Regina Palace in Stresa has an air of faded grandeur

Around Lake Como

Argegno
Locanda Sant'Anna
Sant'Anna, on the Schignano road
Tel: 031-821738; Fax: 031-822046
www.locandasantanna.it
A rural retreat in the hills above Argegno (Como Nord motorway exit), this family-run inn was once a retreat for Como clergy. Eight comfortable, rustic rooms; restaurant overlooking the garden. $

Bellagio
Grand Hotel Villa Serbelloni
Via Roma 1
Tel: 031-950216; Fax: 031-951529
www.villaserbelloni.com
A luxury hotel comparable with Villa d'Este – more personal but less palatial. Facilities are constantly upgraded. Lovely glassed-in restaurant terrace (open in summer); state-of-the-art health spa and beauty treatments. Nightly chamber music; outdoor pool; children's indoor pool with rocks and water slides; private beach, jetty and anchored raft; water sports; tennis. $$$

Firenze
Piazze Mazzini 46
Tel: 031-950342; Fax: 031-951722
Email: hotflore@tin.it
Charming traditional hotel that is a great choice for the discerning traveller. Lakeside terrace; good restaurant and bar; reading room; public rooms feature a sophisticated combination of burnished wood panelling and contemporary furniture; bedrooms are warmly inviting and tastefully furnished; lake views. No grand facilities but the Lido pool is a few minutes away. $$

Cernobbio
Grand Hotel Villa d'Este
Tel: 031-3481; Fax: 031-348844
www.villadeste.it
The undisputed dowager empress of the lake, this sumptuous 16th-century villa has been a hotel since 1873 and features palatial, frescoed rooms studded with antiques. Fans praise the professionalism, critics complain of coldness. Informal grill; formal veranda restaurant; spa; nightclub; floating pool; indoor pool; golf; tennis; sailing. $$$

Como
Le Due Corti
Piazza Vittoria 15
Tel: 031-328111; Fax: 031-265226
This former monastery and post house has been sensitively converted, despite the pool being placed in what were cloisters. The mood is one of quiet good taste, with warm fabrics and exposed beams matched by cosy rooms and a good restaurant. $$

Palace
Lungolago Trieste 16
Tel: 031-303303; Fax: 031-303170
www.palacehotel.it
A luxurious lakeside hotel in the heart of town. Set in what was the archbishop's palace, it has spacious, well-equipped rooms; bar; piano bar. $$

Firenze
Piazza Volta 16
Tel: 031-300333; Fax: 031-300101
www.albergofirenze.it
A well-managed, comfortable 3-star hotel set in the heart of town, on a lively square lined with bars and restaurants. $

Lenno
Albergo Lenno
Via Lomazzi 23
Tel: 0344-57051; Fax: 0344-57055
www.lennoonline.com
Modern 4-star hotel, full of contemporary art, that enjoys a tranquil lakeside setting. Comfortable rooms with whirlpool baths; pool; solarium; fitness centre; restaurant. $$

Menaggio
Grand Hotel Menaggio
Via IV Novembre
Tel: 0344-30640; Fax: 0344-30619
www.grandhotelmenaggio.com
A 4-star lakeside hotel with panoramic views and well-equipped bedrooms. Terrace; gardens; piano bar; gym; heated pool; private jetty. Open Mar–Oct. $$–$$$

Vecchia Menaggio
Via al Lago 13
Tel: 0344-32082; Fax: 0344-30141
Friendly hotel with basic rooms and a restaurant serving good pizza and pasta dishes. $

Tremezzo
Grand Hotel Tremezzo
Via Regina 8
Tel: 0344-42491; Fax: 0344-40201
www.grandhoteltremezzo.com
Set beside Villa Carlotta, this is one of the lake's finest hotels. It offers tasteful period rooms, terraced grounds dotted with modern sculptures. Two restaurants; pool; gym; tennis courts; private jetty; shuttle service. It is a pleasant walk to Bellagio from here. Mar–Nov. $$$

Hotel Villa Marie
Via Regina 30
Tel/fax: 0344-40427
This is a charming, 3-star, newly restored family-run garden hotel overlooking the lake. Appealing terrace, restaurant, small swimming pool and recently renovated bedrooms. Apr–Oct. $$

Around Lake Iseo
Iseo/Clusane
I Due Roccoli
Colline di Iseo, Polaveno
Tel: 030-9822977; Fax: 030-822980
www.idueroccoli.com
A delightful old villa hotel set in the hills 4km (2 miles) from Iseo. Low-key elegance is the keynote, with terracotta floors, exposed fireplaces and airy rooms; pool; tennis. The noted restaurant *(tel: 030-9822978)* features panoramic views, and its specialities include fresh lake produce, home-made salami and fine Franciacorta wines. $$

Iseo Lago Hotel
Via Colombera 2, Iseo
Tel: 030-98891; Fax: 030-9889299
www.iseolagohotel.it
Convenient, well-equipped hotel with short-let service apartments, suites and rooms. Lacks character and lake views but is ideal for families: pool; restaurant and footbridge link to sport and beach facilities. $$

Punta dell'Est
Via Ponta 163, Clusane
Tel: 030-989060; Fax: 9829135
An old-fashioned family-run hotel beside the fishing hamlet of Clusane; the fish restaurant is *the* place to eat baked tench, the local speciality. Rooms are unremarkable but enhanced by lake views and the peaceful sound of water gently lapping the shore. $

Relais Mirabella
Via Mirabella 34, Clusane
Tel: 030-9898051; Fax: 030-989052
www.relaismirabella.it
Set in the hills above Clusane, this 4-star country hotel, converted from an old farmhouse, enjoys panoramic views down to Lake Iseo. Rooms are personalised, light and stylish; pool; landscaped gardens; innovative restaurant and terrace. $$

Franciacorta
L'Albereta
Via Vittorio Emanuele II, Erbusco
Tel: 030-7760550; Fax: 030-7760573
www.albereta.it
A villa hotel designed to showcase the skills of Italy's best-known chef, Gualtiero Marchese. Set in Franciacorta wine country, it is marble-studded and gilded, exuding a spurious glamour. Pool; tennis; spa centre; wine-tasting tours. $$$

Villa Gradoni
Frazione Villa, Monticelli Brusati
(8km/5 miles from Iseo)
Tel: 030-652329; Fax: 030-6852305
www.villa-franciacorta.it
A rural *agriturismo* complex run by the Villa wine-growing estate. The farm buildings have been converted into roomy, comfortable, self-contained flats with exposed beams and stonework; pool; wine-tasting. $

Left: the Palace Hotel in Como is housed in the archbishop's palace

Bergamo
Agnello d'Oro
Via Gombito 22
Tel: 035-249883; Fax: 035-235612
This is an appealing choice in the Città Alta, dating from the 17th century. Modernised smallish rooms and a noted restaurant for regional cooking – stews, polenta, etc. $ (hotel); $$ (restaurant).

Around Lake Garda
Bardolino
Parc Hotel Gritti
Lungolago Cipriani
Tel: 045-6210333: Fax: 045-6210313
This modern well-run waterfront hotel situated on the edge of the historic centre is popular with young families and not far from the ferry stop. Two swimming pools; evening buffets; fitness complex with beauty treatments and massage. $$

Comano
Hotel Villa Luti
Campo Lomaso, Terme di Comano
Tel: 0465-702061; Fax: 0465-702410
This gracious Renaissance country villa 30km (19 miles) from Trento was formerly a patrician residence. Restaurant, bar, tennis, sauna, solarium, Jacuzzi, gym, beauty farm. Shuttle bus to Comano Terme, with its spa treatments and thermal cures. $–$$

Desenzano del Garda
Piccola Vela
Via dal Molin 36
Tel/fax: 030-9914666
Appealing 3-star hotel in olive groves near the centre. Comfortable. $$

Palazzo Arzaga
Cavalgese della Riviera
Tel:030-680600; Fax: 030-6806270
www.palazzoarzaga.com
Luxury 16th-century villa 10km (6 miles) north of Desenzano. Spa and beauty centre; pools; tennis; golf; cycling and walking paths in the grounds; several restaurants. $$$

Gardone Riviera
Grand Hotel Fasano e Villa Principe
Corso Zanardelli 190
Tel: 0365-290220; Fax: 0365- 290221
www.grand-hotel-fasano.it
Once a hunting lodge, this sleek 4-star hotel has lush grounds and a private beach; large differentials between rooms. $$–$$$

Villa del Sogno
Corso Zanardelli 107
Tel: 0365-290181; Fax: 0365-290230
www.gardalake.it/villadelsogno
Elegant Victorian-style villa hotel on a hill overlooking the lake; grand public rooms; spacious bedrooms, charming terrace; pool; solarium; tennis; lakeside beach nearby; short walk to Fasano village. $$–$$$

Villa Fiordaliso
Corso Zanardelli 132
Tel: 0365-20158; Fax: 0365-290011
www.villafiordaliso.com
Exclusive villa hotel linked with Mussolini. Gardens lead to the lake; rooms with a view; top restaurant (*see Eating Out, page 79*). $$$

Limone sul Garda
Le Palme
Via Porto 36
Tel: 0365-954681; Fax: 0365-954120
Email: lepalme@sunhotels.it
Set beside the jetty, this 4-star, 17th-century waterfront hotel is in a quaint but touristy resort. Pleasant rooms; small pool; rooftop sun terrace; lemon grove; a lakeside beach close by; tennis court at sister hotel. $–$$

Above: the Grand Hotel in Tremezzo is often used as a film location

practical information

Malcesine
Parc Hotel Don Pedro
Via IV Novembre
Tel: 045-7400383; Fax: 045-7401100
www.parchotels.it
Self-contained and popular with families, this pleasant 4-star hotel is set in extensive grounds above the resort (a shuttle service goes to the port). Two pools; poolside bar; two restaurants; buffet meals; tennis; mini-golf, volleyball; evening entertainment. $$

Astoria
Tel: 045-7400190; Fax: 045-7400398
www.h-astoria.com
Set on the waterfront 3km (2 miles) from Malcesine (regular day bus service), with pool, terrace and quiet shingle beach. $$

Malcesine
Piazza Pallone
Tel: 045-7400173; Fax: 045-6570073
This traditional, 3-star waterfront hotel is a good touring base. Unexceptional rooms and restaurant; bar with lakeside terrace. $$

Riva del Garda
Hotel du Lac et du Parc
Viale Rovereto 44
Tel: 0464-551500; Fax: 0464-555200
www.hoteldulac-riva.it
This well-equipped 4-star hotel is the best in town but there is little of the style that made it popular with 19th-century writers. Restaurant, bars, beauty centre, gym, two pools, tennis; lush grounds; Closed Nov, Mar. $$$

Hotel Sole
Piazza III Novembre 35
Tel: 0464-552686; Fax: 0464-552811
Email: info@hotel.sole.net
Historic, over-modernised 4-star hotel on the waterfront. Quiet, comfortable rooms; two restaurants; bar; sauna; solarium; close to shingle beach. Closed Nov–March. $–$$

Europa
Piazza Catena 9
Tel: 0464-555433; Fax: 0464-521777
Email: europa@rivadelgarda.com
Central 3-star hotel in modernised palace with views over the port; rooftop terrace with restaurant. Closed Jan–Feb. $$

Villa Moretti
Varone, 3km (2 miles) from Riva
Tel: 0464-521127; Fax: 0464-559098
Family-run hotel set in the hinterland, though the waterfront is accessible by car. Simple rooms, pool and pleasant garden. $

Salò
Laurin
Viale Landi 9
Tel: 0365-22022; Fax: 0365-22382
Delightful Art Nouveau villa in lovely grounds overlooking the lake. Gourmet meals in fine dining room; pool; beach. $$$

Sirmione
Palace Villa Cortine
Via Grotte 12
Tel: 030-9905890; Fax:030-916390
www.hotelvillacortine.com
A romantic deluxe hotel set in parkland a 10-minute walk away from Sirmione. The neoclassical villa has landscaped gardens with fountains, pool, tennis, water sports; private jetty and a path to a beach restaurant. Closed Oct–March. $$$

Hotel Olivi
Via San Pietro 5
Tel: 030-9905365; Fax: 030-916472
www.hotelolivi.it
Secluded 4-star hotel near the Grottoes of Catullus a 10-minute walk from the town centre. Set in an olive grove, this modern hotel has a pleasant garden, pool, fitness centre and restaurant. $$

Spiazzo Rendena
Mezzosoldo
Via Nazionale 196
Tel: 0465-801067; Fax: 0465-801078
Michelin-starred, family-run inn north of Lake Garda. Homely mood; small, cosy rooms; games room and garden. $

Torbole
Lido Blu
Via Foci del Sarca 1
Tel: 0464-505180; Fax: 0464-505931
Email: lidoblu@lidoblu.it
Modern 4-star hotel set on a spit of land in the water. Fine facilities: private beach, pool, gym, windsurfing school. $$–$$$

EMERGENCIES

Medical (Ambulance), tel: 118
All-night pharmacies (Italy-wide service), tel: freephone 800-801185;
Milan 24-hour pharmacy, tel: 02-6690735
Military Police *(Carabinieri)*, tel: 112
State Police, tel: 113
Fire Brigade, tel: 115
ACI (Italian Automobile Association, for breakdowns), tel: 116
Lake Garda coastguards, tel: freephone 800-484848.

Security & Crime
The lakeside resorts are generally very safe, as are most of the small towns, although women travelling alone run the risk of being hassled in the evening. In Milan, a watchful eye should be kept on possessions, especially around the Central Station. In Brescia there is little sense of tourism, and an undercurrent of racial tension, but the places suggested in the itinerary are generally fine.

Medical
In theory, EU and Australian citizens are entitled to reciprocal care, but you should nonetheless get travel insurance. In a real emergency, you will probably be seen promptly and efficiently at a local hospital. Pharmacies *(farmacie,* generally 8am–noon, and 4–7pm) are a good first stop for medical advice. The name of the duty or all-night pharmacy is posted on pharmacy doors.

COMMUNICATION & NEWS

Postal
Posta prioritaria is more expensive than the regular service but is the fastest, most reliable way by which to send post.

Telephone, Fax & Email
The Italy country code is 39. If phoning any location in Italy, the full area code is required. The region has relatively few public phones. Though some take cash, the majority of public phones either take phone cards *(schede telefoniche)*, which can be bought from tobacconists *(tabacchi)*, or credit cards. Many services and attractions now have freephone numbers (which begin with 800) and need no other code, but these can only be dialled within Italy. Although Italians are enthusiastic about the internet, and cyber-cafés are popular in bigger cities, email communications still lag behind the UK and the US.

Media
The main newspapers in the north are the influential Milan-based, centre-right daily *Corriere della Sera*, and its centre-left rival, *La Repubblica*. Both contain listings for entertainment in Milan. There are three main public TV channels and many private ones.

USEFUL ADDRESSES

Italian tourist offices abroad
UK: ENIT (Italian State Tourist Board): for all regions in the Lakes except Trentino: 1 Princes St, London WLR 8AY; tel: 020-7408 1254; fax: 020-7493 6695; www.enit.it.
Trentino Information Service, tel/fax: 020-8879 1405; www.trentino.to; email: trentino.infoservice@virgin.net.
USA: 630 5th Avenue, Suite 1565, New York, NY 10111; tel: 212-245 5618; fax: 212-586 9249; www.italiantourism.com.
Canada: 17 Bloor St East, Suite 907, South Tower, Toronto, Ontario M4W 3R8; tel: 416-925 4882; fax: 416-925 4799; www.italiantourism.com.

Tourist offices in the lakes
APT Milan: Via Marconi 1 (by the Duomo). Tel: 02-725241; fax: 02-72524350); www.milanoinfotourist.com

Around Lake Maggiore:
Lake Maggiore Tourist Board (Distretto dei Laghi), Via Principe Tommaso 70, Stresa. Tel: 0323-30416; fax: 0323-934335; email: infoturismo@distrettolaghi.it.
Stresa: Via P. Tommaso 70. Tel: 0323-30416; fax: 0323-934335.
Verbania–Pallanza: Corso Zanitello 6. Tel: 0323-503249; fax: 0323-556669.
Lake Orta: Via Panoramica, Orta San Giulio. Tel: 0322-905163; email: inforta@distrettolaghi.it.

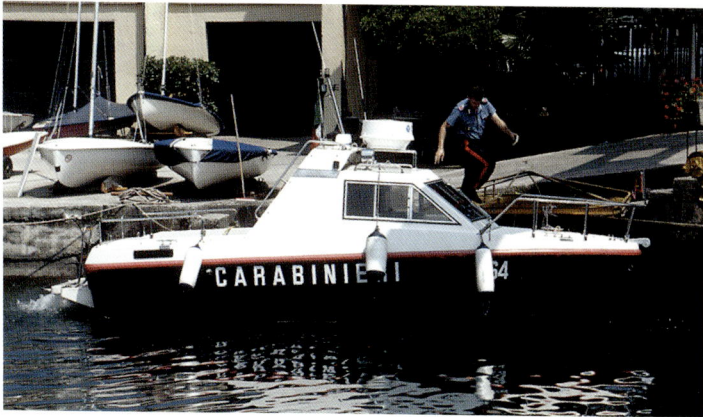

practical information

Around Lake Varese and Lake Lugano:
APT del Varesotto (the province): Tel:
0332-284624; fax: 0332-238093.
Varese (the city) Via Carrobbio 2.
Tel/fax: 0332-283604).
Around Lake Como:
APT Bellagio: Piazza della Chiesa. Tel/fax:
031-950204.
APT Cernobbio: Via Regina 33/b. Tel/fax:
031-510198.
APT Como: Piazza Cavour 17. Tel: 031-
3300111; fax 031-240111; www.lake
como.org
APT Menaggio: Piazza Garibaldi 8. Tel:
0344-32924; fax: 0344-510198.

Around Lake Iseo:
APT Bergamo: Viale V Emanuele 20 (lower
town). Tel: 035-213185; fax: 035-230184;
also Vicolo Aquila Nera 2 (upper town). Tel:
035-232730.
APT Iseo: Via Stazione G Marconi 2. Tel:
030-980209; fax: 030-981361.
APT Pontedilegno: Corso Milano 41. Tel:
0364-91122/903838; fax: 0364-91949.

Around Lake Garda:
APT Brescia: Corso Zanardelli 34. Tel: 030-
43418/45052; fax: 030-293284.
APT Desenzano del Garda: Via Porto Vec-
chio. Tel: 030-9141510 ; fax: 030-9144209.
APT Garda (town): Lungolago Adelaide:
Tel: 045-6270384.
APT Gardone Riviera: Corso Repubblica
35. Tel/fax: 0365-20347.
APT Limone sul Garda: Via Comboni 15.
Tel: 0365-954070; fax: 0365-954689.

Above: water watch
Right: background reading

APT Malcesine: Via Capitanato 6. Tel: 045-
7400555.
APT Riva del Garda: Giardini di Porta Ori-
entale 8. Tel: 0464-554444; fax: 0464-
520308; www.gardatrentino.it.
APT Sirmione: Viale Marconi 2. Tel: 030-
916114/916245; fax: 030-916222.
APT Torbole, Via Lungolago Verona 19. Tel:
0464-505177; fax: 0464-505643.
Trentino Tourist Board: Via Romagnosi 11,
Trento. Tel: 0461-839000; www.trentino.to
Verona: Piazza Brà Via degli Alpini 9.
Tel: 045-8068680; www.tourismverona.it.

FURTHER READING

Insight Guide to Northern Italy. Background
essays, a comprehensive look at the entire
region, colour photographs and maps.
Insight Pocket Guide to Milan. Itineraries
around Milan; complete with pull-out map.
Insight Compact Guide Italian Lakes.
Packed with facts, pictures and maps.

ACKNOWLEDGEMENTS

Photography	
12B	**AKG London**
14T	**AKG London/Cameraphoto**
10	**AKG London/Vision Ars**
15, 16T	**The Art Archive**
1, 2/3, 5, 8/9, 20, 21, 22, 23T&B, 24T&B, 26, 27T&B, 28, 29T&B, 30T&B, 31, 32, 33T&B, 34, 35T&B, 37T&B, 38B, 39, 41T&B, 42, 43, 44T&B, 45, 46T&B, 47, 54B, 72, 74T&B, 76, 80, 83, 86, 89, 93, 95, 96	**Annabel Elston**
71, 73	**Jerry Dennis**
11	**Glyn Genin**
25, 49T&B, 68, 69, 70	**Ros Miller**
53B	**Mark Read**
85	**Rex Features**
12T	**Rex Features/Sipa Press**
84T	**Topham Picturepoint**
84B	**Topham/Empics**
16B, 50T&B, 51, 53T, 54T, 55, 57T&B, 58, 59, 60, 61T&B, 63T&B, 64T&B, 65, 66T&B, 67, 75, 77, 78, 81, 82T&B, 87, 91, 99T&T	**George Taylor**
14B	**Courtesy Villa Cicogna Mozzoni**
38T	**Bill Wassman**
Front Cover	**Charlie Waite/Stone/Getty Images**
Back cover Top	**George Taylor**
Back cover Bottom	**Annabel Elston**
Cartography	**Berndtson & Berndtson**

INDEX

i
n
d
e
x